The Early-Career Professional's Guide to Generative AI

Opportunities and Challenges for an AI-Enabled Workforce

Jonas Bjerg

The Early-Career Professional's Guide to Generative AI: Opportunities and Challenges for an AI-Enabled Workforce

Jonas Bjerg
København S, Denmark

ISBN-13 (pbk): 979-8-8688-0455-7   ISBN-13 (electronic): 979-8-8688-0456-4
https://doi.org/10.1007/979-8-8688-0456-4

Managing Director, Apress Media LLC: Welmoed Spahr
Acquisitions Editor: Shivangi Ramachandran
Development Editor: James Markham
Project Manager: Jessica Vakili

Cover designed by eStudioCalamar

Distributed to the book trade worldwide by Apress Media, LLC, 1 New York Plaza, New York, NY 10004, U.S.A. Phone 1-800-SPRINGER, fax (201) 348-4505, e-mail orders-ny@springer-sbm.com, or visit www.springeronline.com. Apress Media, LLC is a California LLC and the sole member (owner) is Springer Science + Business Media Finance Inc (SSBM Finance Inc). SSBM Finance Inc is a **Delaware** corporation.

For information on translations, please e-mail booktranslations@springernature.com; for reprint, paperback, or audio rights, please e-mail bookpermissions@springernature.com.

Apress titles may be purchased in bulk for academic, corporate, or promotional use. eBook versions and licenses are also available for most titles. For more information, reference our Print and eBook Bulk Sales web page at http://www.apress.com/bulk-sales.

Any source code or other supplementary material referenced by the author in this book is available to readers on GitHub (https://github.com/Apress). For more detailed information, please visit https://www.apress.com/gp/services/source-code.

If disposing of this product, please recycle the paper

To my late grandmother, Eva, who taught me how to love books.

To my dad, Jens, for showing me the hard parts of life and for teaching me to dream big.

To my mom, Birgitte, for showing me the soft parts of life and for teaching me to stay grounded.

To my brother, Mathias, for never telling mom and dad what trouble we were up to.

To my sister, Sara, for always telling mom and dad what trouble we were up to.

To my best friend, Kristian, who taught me how to make more friends.

Without you all, I would have never been able to finish this book.

Table of Contents

TABLE OF CONTENTS

About the Author

Jonas Bjerg is currently an Expert Manager, Data Science, at Bain & Company Inc. He manages multiple cross-functional teams of data science specialists. He has previously lived and worked in big tech in Silicon Valley for a few years. He is a thought leader that presents at several conferences including Impuls 2023 and CPH AI HUB, topics that usually cover the future of AI and how businesses can utilize it to build an advantage. He has a successful YouTube channel about technology news that has more than 11,000 subscribers and millions in video views.

Acknowledgments

It does not do well to dwell on dreams and forget to live.

—Albus Dumbledore

It has always been a dream of mine to publish a book one day. The journey to this publication has been an adventure with a steep learning curve. But I can't dwell on it any longer; now, I must return to living.

This book would never have come to be if it weren't for all my great teachers, colleagues, and friends.

Special thanks to

- **Ted Ladd** for teaching me much more than what was on the curriculum.

- **Ronjon Nag** for teaching me that neural networks are more than just math and statistics.

- **Thomas Kurnicki** for teaching and introducing me to natural language processing.

- **Sara Toft Tjell**, **Jens Friis Hjortegaard**, and **Torsten Hvidt** for giving me the dream job I didn't know I needed.

- **Linda Raaijmakers, Agnieszka Turkiewicz, Ivy Intano, Aditya Joshi, Andreas Kaempf**, and **Hesham Mahrous** for being the inspiring, brilliant people you are. You are the best colleagues I could hope for. Thank you for challenging my knowledge every day!

ACKNOWLEDGMENTS

- **Gianluca Mauro** and **Nicolò Valigi** for inspiring me to write this book.

- **Teresa Galli**, **Annemette Møhl**, and **Jashan Sippy** for proofreading my scribbles and for being wonderful people.

- **Shivangi Ramachandran** and **Apress** for seeing the potential in my scribbles.

CHAPTER 1

Introduction

Almost every headline today is about how artificial intelligence (AI) is going to ruin the world. It will take all our jobs and leave us all in the dust as soon as Artificial General Intelligence (AGI) comes along. This narrative bothers me immensely. It bothered me enough to start writing this book. Make no mistake, AGI is indeed coming, and it will never be the same again once it gets here. But the same could be said for the writing press, the Internet, or every other major historical invention. Anyone who claims to know when or how our world will change instantly is lying to themselves and to everyone else. The doomsday narrative currently dominating the headlines helps nobody except those selling newspapers.

Some experts have said that they fear the development of AI, and some have even started advocating for a total ban on all AI research and development. I sympathize with these experts' perspectives. This book is not an attempt to discredit the potential detriment a mean-spirited AGI can have on our world. But I don't believe we can stop the development, not even with a global ban tomorrow. The genie is out of the bottle, so what is next? The newspapers want us all to panic—and buy tomorrow's paper. I agree with the need for oversight and regulation, but I disagree entirely with the panic agenda. Moore's law states that the number of transistors in an integrated circuit doubles every two years. This law has been in effect for decades, since 1965, and lately, it has proven too slow to keep up with the recent surge in processing power improvements, according to Nvidia CEO Jensen Huang at the GTC conference 2023.

© Jonas Bjerg 2024
J. Bjerg, *The Early-Career Professional's Guide to Generative AI*,
https://doi.org/10.1007/979-8-8688-0456-4_1

Regarding chip manufacturing, Moore's law is struggling to keep up nowadays, presumably due to the rise in demand from the AI industry. The fact that something will be better in the future is almost a universal truth regarding technology. Very few things decrease in technical capability over time. Total bans on nuclear technology or illegal drug development have done very little to slow their technological progress. So, if we accept that AI is here to stay, what's next? And how should we prepare for such a future?

I'm writing this book primarily to help students because I believe this doomsday narrative hurts them the most. Students today have to pick a profession, spend a little decade mastering it, and hopefully start making a living off it. This takes dedication and willpower. And it only gets harder if everyone around you keeps telling you that your job will be replaced tomorrow, so why even bother? This word "soon" also bothers me. They don't know. The experts don't even know. Five to ten years ago, all experts agreed that it simply was a given that creative jobs would be the hardest for an AI to replace. This prediction was made almost unanimously by all industry experts at the time. They didn't know how a simple transformer model could impact the world. The ironic part is that the transformer model that proved capable of these creative tasks was first published in a paper in 2017, but no expert managed to put two and two together in time to correct their prediction. Granted, the fact that experts made this statement probably also impacted investors' willingness to throw larger amounts of money toward solving these "most difficult AI problems," but the statements were proven entirely wrong, and creative jobs were, in fact, some of the first to be threatened by AI—threatened even before all the other "repetitive tasks," which were supposed to be the easiest ones to replace. Even self-driving cars have been projected to be solved by 2020— that didn't happen, at least not for level 4 autonomy, but we are very close. If the experts can't accurately predict which areas will be automated first, how is anyone else supposed to? The concept of "soon" can change with a single discovery. Superconductors at room temperature were thought to be a science fiction thing of the future, but scientists from Korea University in

Seoul got incredibly close recently. It didn't turn out to be the final solution to superconducting we were hoping for. Still, the Korean learnings will undoubtedly help the next scientist get even closer—turning the "soon" into a "now" sooner than any expert can predict.

Welcome to *The Early-Career Professional's Guide to Generative AI*. My name is Jonas Bjerg, your frustrated narrator, who will do his utmost to explain the world of artificial intelligence (AI) in as simple terms as I can manage. Even if you don't know anything today, this book will show you all the interesting nooks and crannies of AI as it stands today and give you an overview of the fundamentals of AI as well as its potential impacts on the world, as I see them—expecting to be proven wrong on several counts, like countless experts before me.

Before we delve into AI's complexities and opportunities, let's start with the classic, "Who am I, and why am I qualified to even talk about these topics?"

I've had the privilege of studying at two of the world's premier institutions. I was fortunate enough to study under some of the best minds in the field; their insights and cutting-edge curriculum shaped my understanding of AI and its potential—it laid my foundation.

I've studied law, international business, and business analytics (the core data science part). After graduating with top marks and all the bells and whistles I could dream of, I entered the tech arena, landing a coveted position at one of Silicon Valley's top tech firms. Here, I was exposed to the rapid advancements in technology and the practical applications of AI. This hands-on experience and my academic background gave me a holistic view of AI's capabilities and shortcomings.

I am wearing the hat of a data science expert manager at Bain & Company, a globally renowned consulting firm. In this role, I've consulted for some of the world's most influential companies, advising them on advanced analytics and the transformative power of Neural Networks and AI in general. If I were to put a holistic tagline to my current job, it would be as pretentious as, "We help people do the impossible," and the secret sauce that allows us to do it is almost always statistics and AI—from building and deploying countless neural networks to tackling unconventional problems that haven't been solved before. One of my more memorable projects involved predicting characters' fates in the final season of "Game of Thrones"—a blend of pop culture and advanced technology! Another project that stands out is developing a model to predict the likelihood of product returns at the point of purchase. It offers businesses invaluable insights to optimize their operations and lower their emissions.

Beyond my work, I've become somewhat of an AI evangelist, speaking at various international forums and conferences where I am sharing my knowledge and insights on AI's impact on society.

I've engaged in enlightening debates with thought leaders, industry experts, and curious minds worldwide.

I cringe rather harshly writing down these "accomplishments," but I think they are important for you, the reader, to know before we proceed.

Now, back to the regularly scheduled programming. As I pen down my thoughts in this book, the world stands at a pivotal moment. In recent years, these moments seem ever more frequent. This time, it is because of the emergence of ChatGPT from OpenAI. This groundbreaking technology has provoked and impressed almost every industry—evoking every emotion from awe to anxiety. The headlines provide no solace here. Many are apprehensive about the future, fearing job losses due to rapid AI advancements. But if history has taught us anything, progress, while challenging, often paves the way for broader opportunities and growth. This is not apparent if one only reads the recent news articles. Through this book, I aim to share my convictions, insights, and vision for an AI-integrated future, hoping to provide clarity and optimism in these exciting yet uncertain times.

First, I will explain the core building blocks that make up the current landscape of transformer models and, more broadly speaking, AI as a whole. Once a basic understanding of these concepts is clear, we can start the philosophical discussions and bring everything in a historic and societal context to help calm the doomsday-sayers. It is not the first time a new technology has significantly changed our world, and it won't be the last. So, let's learn from the past and act accordingly in an informed way.

The latest significant breakthrough in mainstream media is ChatGPT. And it truly is a remarkable achievement in the realm of artificial intelligence. But what is it actually? The "GPT" in its name has coined quite a number of phrases in pop culture, but what does it stand for? "Generative Pretrained Transformer." To break it down:

- **Generative:** This term indicates an ability to generate output, often in the form of text. Rather than just analyzing or processing information, it can produce entirely new content, often with a level of coherency and creativity that rivals human output.

- **Pretrained:** This means that the model is "fine-tuned" for specific tasks. Essentially the model undergoes extensive training on vast amounts of data. This pre-training phase equips the model with a broad understanding of the broader contexts, for example, language, syntax, etc., enabling it to tackle various tasks immediately, without first requiring to train on additional data from scratch.

- **Transformer:** This refers to the model's architecture or framework. This secret sauce that made ChatGPT so much better than previous models. The Transformer architecture revolutionized the field of deep learning, especially for tasks involving context. It allows models to pay selective attention to different parts of the input data, thereby understanding context and relationships more effectively.

GPT models started relatively simple. They were designed primarily to predict the subsequent word in a sequence. Imagine typing a sentence on a phone and having the system suggest the next possible words in the top bar of the keyboard—that's the initial capability of these models.

However, as technology advanced and more data became available, GPTs became useful in a broader array of use cases. From its humble beginnings, it grew in both size and sophistication. The models not only became adept at predicting words but also began to grasp context, discern nuances, and, to a certain extent, capture emotional undertones in text.

The world took notice when the third iteration, ChatGPT, was released. This wasn't just another incremental improvement; it represented a quantum leap in AI's ability to understand and generate human-like text. The sheer scale, fine-tuning, and advanced techniques made ChatGPT a sensation in the tech world and beyond.

I asked ChatGPT to explain how its predecessor was trained and what it can do—note that it talks in first-person:

"My training involved learning from a diverse range of internet text. However, while I strive to provide accurate and helpful information, I can only access personal data about individuals if shared during our conversation. I am designed to respect user privacy and confidentiality.

Over the years, I have been utilised in various applications, from drafting emails and writing code to creating written content and tutoring in multiple subjects. My ability to generate human-like text has made me a valuable tool for many businesses and individuals worldwide."—ChatGPT

While ChatGPT has gained significant attention for certain applications, it's essential to recognize that its capabilities extend far beyond what's immediately evident. Artificial Intelligence, represented by models like ChatGPT, is not a static field. It's dynamic and ever-evolving, and its potential applications are broadening each day. I have, for instance, built a video editor that automatically cuts out errors and removes stutters, silences, etc., using language models similar to ChatGPT's models. This is to say that the technology is not, by any means, limited to chatbots or translation use cases. This book endeavors to capture this vast and ever-expanding horizon of possibilities.

In the next chapters, we will discuss language models in depth. These are not just algorithms; they represent a nexus of linguistics, cognitive science, and cutting-edge technology. We'll trace AI's unexpected and exhilarating evolution, observing how it has grown from a mere concept to a force reshaping entire industries.

For entrepreneurs and business leaders, the emergence of AI goes beyond being a trend. It presents a plethora of opportunities waiting to be harnessed. I will give examples of ways businesses can leverage AI to capitalize on these opportunities in later chapters.

As I mentioned earlier, only some people share this positive outlook on the future. Some challenges and implications accompany the rise of AI. One major concern on people's minds is the displacement of jobs. Will AI result in mass layoffs? If so, how should society adapt? We will discuss solutions, including concepts like Universal Basic Income (UBI) and Artificial General Intelligence (AGI). Can UBI serve as a safety net that ensures a balance between progress and societal well-being when AGI takes the stage? Without getting too political, I will try to highlight the positives and the negatives of such a society in later chapters.

Art and creativity were once considered impervious to AGI, and now even a simple language model has proven it can challenge jobs in this arena and generate stunning images with ease. But these outputs fall short compared to an AGI model's theoretical, artistic potential. I guarantee you that even a top-notch language model will fade in comparison. But what does this mean for society? If we are to listen to the news then it is all pointless because the AGI will come to take our jobs soon anyways. I simply refuse to subscribe to this way of thinking. If an artist spends 1000 hours obsessing over a piece, putting sweat and tears on the canvas, that final product will always be worth more to me, than an AI generated identical replica. So is there a world where both AGI and artists can co-exist? I believe we already live in such a world now. The invention of digital images didn't do much to stifle the art scene, quite on the contrary, it became a new tool for artists to speed up their process—especially in regards to prototyping concepts, a role AI also could be immensely helpful in.

In this book's later chapters, we will engage in a thought-provoking discussion about artworks generated by AI and the artwork's legitimacy in the art scene. How will the art world value these generated pieces? Does art created through algorithms possess the ability to be trademarked or copyrighted? Does a sad emotional piece lose its depth when you realize its creator never understood or felt the potential suffering it portrays? Ask yourself these questions. Would you pay the same for a piece generated in 20 seconds by a machine if it looks identical to the piece created over many late nights by a real artist? I expect many of us will have differing opinions. Still, considering similar historical situations, we already know what the average person would prefer. When the printing press came out, the price of a book decreased immensely; the same was true when the digital printer came out.

Suddenly, a poster was a thing and something new you could hang on your walls. Did these inventions mean that artistry disappeared entirely? No, but they did affect the price of a digital painting. Now, AI will most likely do the same again. Some call it democratizing art; others call it plagiarism. And both are right in their own way. An obscene amount of real art is needed to train these art generators as a training dataset. It will be interesting to see how the legal aspects pan out. As recent lawsuits have already revealed, only some companies have had the same approach to legally obtaining their datasets. On August 21, 2023, a US court in Washington, D.C., ruled that a work of art created by artificial intelligence without human input cannot be copyrighted under US law. The ruling surprised me; remember that I only have a bachelor's in Danish law, but from a legal perspective, the US has been quite firm on giving companies the same rights as people, as shown in the Supreme Court case Roe v. Wade. But this ruling on AI art showed that the judge wanted a human involved in the creation before they would grant copyrights on the output. This is, of course, a decision favoring the individual artists, which I fully support. Still, I expected them to show similar favor toward the machines as they had the corporations in the past.

Lastly, it's important to consider the rise of AI in the context of advancements. We have seen how the Internet and the mobile revolution changed our world. Is AI following a similar trajectory? Are we on the verge of something even more transformative? Elon Musk and other industry leaders believe that we are. They signed a letter urging lawmakers to pause AI research for months to take action. Elon's main concern is that regulations usually come into effect after something has gone wrong, but with AGI (Artificial General Intelligence), it might be too late. Interestingly, Elon was instrumental in launching OpenAI as a non-profit organization, he has since left the company and is now actively advocating against it, saying it is a for profit company essentially owned by Microsoft (because they are the largest shareholder in OpenAI). After his departure, Elon publicly and loudly stated that he supports regulations against AI, stating that he fears this technology more than any other threat to our world. I believe he is right that this technology holds immense potential and that it needs to be kept in check. Both in terms of bad actors (humans with bad intentions) and in terms of avoiding catastrophes. The question comes down to how, because I don't think even a global ban can stop the development. In my humble opinion, the genie is out of the bottle at this point. Elon predicts that AGI will be built in the next decade, I don't have quite the same view. I see us being quite a bit further away from that release date, but we can definitely agree that it is coming "soon."

Join me on this journey as we explore the landscape of AI, delving into its marvels and addressing its challenges.

In this changing world of AI, I strive to share my insights and experiences to shed light on all the interesting bits and pieces of this field, as well as their potential and shortcomings.

My main goal is to bring clarity to the state of AI, explain its impact, and offer guidance on navigating this ever-changing field confidently. I aim to assist students currently pursuing subjects that may be affected by automation or replaced by language models in the future. I fully understand the discouragement that can arise from the possibility of industry shifts for those who have worked hard to enter their chosen field. The media and experts often focus heavily on the changes that lie ahead. However, history has shown us a pattern. In the past, we have experienced shifts in platforms that could have left workers jobless within their industries. Instead, what typically has happened is that industries adapt and embrace technologies such as automation, robotics, computers, the Internet, or mobile devices, resulting in increased productivity and output. It's important to consider networks and language models before prematurely dampening students' motivation to study specific subjects. Take knee surgeons as an example; they undergo exceptional amounts of education.

However, the idea of a robot surpassing surgeons in performing knee surgery is plausible in the near future. It will likely become a reality, posing challenges for medical students aspiring to specialize in this field. Undoubtedly, there will be a phase where people will trust doctors more than robots, but this sentiment won't persist indefinitely. The mere statistics will persuade the public and their trust will start to lean toward the statistically safest choice. Once the transition is complete, we will never revert to relying on surgeons for such procedures, much like how we no longer fly planes without autopilot or navigate using a physical map when driving. However, this raises questions about the role of surgeons in the future. Does it imply that aspiring doctors should not pursue this profession anymore? If it takes ten years to become a surgeon and Elon predicts that AGI's that can do everything are coming in the next ten years, then how will I find a job as a surgeon? This is a very logical conclusion to come to as a student, but I urge you to stop thinking this way. Elon, everyone else, and myself are not able to tell the future, so don't base your life on our estimates. Surgeons will continue to be crucial for years to come. So if your dream is to become a surgeon, then do everything you can to become one!

Moreover, it's uncertain which surgeries will be automated first. Even if your specific area of expertise becomes automated, having a medical degree remains invaluable. You could explore options such as teaching, re-educating yourself in another specialization or even transitioning to another department entirely. Throughout history, similar technological disruptions have occurred. Individuals have adapted accordingly and many have still found ways to be employable in their profession. It is likely that you won't be the one performing the surgery with your own hands, but that doesn't mean you won't be needed in a world where the robot does the cutting and you do the human job of explaining the procedure and perhaps overseeing the procedure on standby. There are a hundred ways this could play out where your job as a surgeon is "saved," so don't listen to the naysayers—follow your dreams!

Fortunately, history provides reassurance that human expertise remains relevant and indispensable despite advancements and disruptions. Even if there comes a time when a specific surgical procedure becomes automated, the value of obtaining a degree will remain significant. Throughout history, humans have consistently demonstrated their resilience and adaptability whenever technology threatens roles.

In essence, the message I want to convey is one of optimism. The emergence of neural networks and advanced language models should not demotivate us. Instead, we should see them as tools that enhance our capabilities. While automation will undoubtedly play a role across various industries, the necessity for human expertise, judgment, and intuition will endure. So instead of fearing the surge of AI technology, let's embrace it. Let us harness its power, adapt to the changes it brings forth, and mold it to meet our goals. The future may be uncertain. It is full of opportunities for those who are open to evolving.

So don't fear AI; utilize it to your advantage. Adjust accordingly. Experts in these fields will continue to be needed for a long while. So you won't lose your job to an AI right away, but you might lose it to someone who has embraced it and uses it to "work smarter."

Now, let us embark on this learning journey together as we explore the captivating world of AI and its implications for our future.

CHAPTER 2

A Brief AI History

I am a big believer in learning from the past to form our estimates for the future; to allow you the best possible introduction to this field without going into too many details, this chapter is very short and to the point. It essentially highlights the relevant people and eras in AI development before the introduction of the language model.

The Evolution of AI

The journey of AI from a theoretical concept to a practical, impactful technology is a fascinating one. The evolution of AI can be broadly segmented into three phases:

Early AI (1950s–1970s)

The early years of AI were characterized by groundbreaking theoretical work, significant research efforts, and a general sense of optimism about the potential of this nascent field. This period laid the foundation for modern AI and witnessed the emergence of some of the most influential ideas and technologies that still underpin the field today.

Pioneers of the Theoretical Foundations

Alan Turing and the Turing Test: Alan Turing, often regarded as the father of computer science and AI, made one of the most significant

© Jonas Bjerg 2024
J. Bjerg, *The Early-Career Professional's Guide to Generative AI*,
https://doi.org/10.1007/979-8-8688-0456-4_2

contributions to the field. His 1950 paper, "Computing Machinery and Intelligence," proposed what is now known as the Turing Test—a method for determining whether a machine is capable of intelligent behavior equivalent to, or indistinguishable from, that of a human. This test set a benchmark that continues to influence AI development.

John McCarthy and the Dartmouth Conference: In 1956, John McCarthy, often credited with coining the term "Artificial Intelligence," organized the Dartmouth Conference. This pivotal event brought together researchers interested in neural networks, language processing, and the study of intelligence, effectively marking the birth of AI as a field of study.

Technological Advances and Early Successes

Rule-Based Systems: One of the earliest successes in AI was the development of rule-based systems, where machines used sets of predetermined rules to solve problems and make decisions. These systems demonstrated that machines could exhibit aspects of intelligent behavior.

ELIZA—The First Chatbot: In the mid-1960s, Joseph Weizenbaum created ELIZA, one of the first chatbots. ELIZA used a pattern-matching and substitution methodology to simulate conversation and was able to create an illusion of understanding, though it had no built-in framework for contextualizing events.

Chess-Playing Programs and Decision-Making: AI research in this era also saw the development of chess-playing programs. These programs, while primitive compared to today's standards, were crucial in advancing techniques for decision-making and strategic planning in AI.

Challenges and Limitations

Despite these successes, the early years of AI were not without challenges. The limited computational power of the time was a significant barrier, restricting the complexity of tasks that AI systems could handle. Moreover, there was a growing realization that simulating human intelligence was more complex than initially anticipated, involving not just logical reasoning but also learning, perception, and emotional intelligence.

Legacy of the Early AI Era

The early period of AI set the stage for all subsequent development in the field. The pioneering work of early researchers not only provided vital theoretical foundations and technological advancements but also sparked a wave of interest and investment in AI that would grow over the decades. This era, marked by optimism and groundbreaking research, established the bedrock upon which the diverse and complex field of AI would be built.

AI Winter (1980s–1990s): A Period of Reevaluation and Quiet Progress

The AI Winter refers to a period during the 1980s and 1990s when the initial enthusiasm for Artificial Intelligence cooled significantly. This phase was marked by reduced public interest, waning funding, and a general sense of skepticism toward the field. The high expectations set during AI's nascent years had not been met, leading to disappointment among researchers, investors, and the public.

Factors Leading to the AI Winter

- **Over-promised and Under-delivered:** In the early years of AI, there was a tendency to over-promise the capabilities of AI systems. When these systems failed to deliver the revolutionary changes that were anticipated, it led to a loss of credibility and disappointment.

- **Limitations of Technology:** The technological limitations of the time, particularly in terms of computational power and data storage, restricted the complexity and effectiveness of AI applications, leading to disillusionment.

- **Lack of Sustained Funding:** As a result of these unmet
  expectations, both government and private funding for
  AI research dwindled. The reduction in funding further
  hampered the development of the field.

The Silver Lining: Groundwork for Future Advancements

Despite the challenges, the AI Winter was not devoid of progress.
In fact, this period was critical in laying the foundation for future AI
advancements:

Advances in Neural Networks: During this time, researchers continued
to work on neural networks, though much of this work went unnoticed by
the broader public. Key theoretical advances were made, including the
development of backpropagation algorithms, which later played a vital role
in training deep neural networks.

Focus on Specific Applications: The AI community began focusing on
more achievable, specific applications rather than attempting to create
general AI. This pragmatic approach led to successful implementations in
areas like logistics, manufacturing, and data mining.

Development of Machine Learning: The period saw the refinement of
machine learning techniques. These techniques, especially in areas like
decision tree learning and reinforcement learning, would later become
central to the AI resurgence.

Setting the Stage for a Resurgence

The AI Winter, while a period of reduced enthusiasm and funding, was
crucial for recalibrating the expectations and approaches in AI research.
It was a time of introspection and consolidation for the AI community,
which allowed for a more realistic and focused approach to research
and application. The groundwork laid during this period set the stage for

the subsequent resurgence of AI, driven by advances in computational power, data availability, and algorithmic innovation in the late 1990s and early 2000s.

Modern AI Renaissance (2000s–Present): The Era of AI Revolution

The Modern AI Renaissance marks a period where Artificial Intelligence transitioned from theoretical and experimental phases to becoming a dominant force in the real world. This era, beginning in the early 2000s and continuing to the present, is characterized by rapid advancements, widespread adoption, and a significant impact on various aspects of society and industry.

Key Drivers of the AI Renaissance

Explosion of Data: The digital age has led to an unprecedented increase in data generation. From social media interactions to IoT (Internet of Things) devices, data is being produced at an astonishing rate. This wealth of data has become the fuel for AI, providing the raw material for machine learning algorithms to learn and improve.

Advancements in Computational Power: Significant strides in hardware, such as GPUs (Graphics Processing Units) and TPUs (Tensor Processing Units), have provided the necessary computational power to process large datasets and run complex algorithms. This has allowed for the training of more sophisticated AI models, including deep learning networks.

Innovations in Machine Learning Algorithms: There have been major breakthroughs in machine learning algorithms, particularly in deep learning. Neural networks, especially Convolutional Neural Networks (CNNs) for image recognition and Recurrent Neural Networks (RNNs) for sequence analysis, have achieved remarkable success.

Impact and Applications

The impact of the modern AI renaissance is far-reaching and transformative:

- **Transformation of Industries:** From health care, where AI is used for diagnostics and personalized medicine, to finance, where it's applied in algorithmic trading and risk management, AI is revolutionizing industries.

- **Advancements in Autonomous Vehicles:** AI has been a driving force behind the development of autonomous vehicles, promising to reshape transportation.

- **Revolution in Communication:** AI-powered natural language processing has led to more effective and human-like chatbots and virtual assistants, changing how we interact with technology.

- **Enhancements in Personalization:** Whether in online shopping, content streaming, or digital marketing, AI's ability to analyze user preferences has led to unprecedented levels of personalization.

Challenges and Ethical Considerations

Despite its successes, the modern AI era faces its own set of challenges:

- **Ethical Concerns:** Issues such as algorithmic bias, privacy, and the ethical use of AI have sparked significant debate.

- **AI and Employment:** The impact of AI on job markets and the potential displacement of certain types of work have become areas of concern and research.

- **AI Governance:** As AI becomes more influential, the need for regulations and guidelines to ensure its responsible and ethical use becomes increasingly important.

Conclusion

The AI renaissance continues to evolve, driven by ongoing research, increasing investment, and a growing understanding of AI's potential and challenges. The future of AI promises even more integration into everyday life, with potential breakthroughs in areas like AI-driven healthcare, sustainable energy solutions, and more advanced human–computer interactions.

The growth of AI is a testament to human ingenuity and its potential to continue reshaping our world is immense. As we stand on the cusp of AI-driven transformation, understanding its potential and responsibly harnessing its power is more important than ever.

CHAPTER 3

Understanding Language Models

Language models have a role in natural language processing (NLP), a branch of artificial intelligence that focuses on the interaction between computers and human language. In essence, language models are structures created to predict the likelihood of word sequences occurring in a given language. This predictive ability forms the basis for applications ranging from spell-check and autocomplete features to complex tasks like machine translation and text generation.

In this chapter, we will delve deeper into how these language models work, explore the networks that power them, and trace their evolution. We will also explore the applications of language models, their limitations, and what the future holds for this field of AI.

A Closer Look...

Language models have roots in linguistics, where researchers have long been studying the structure and rules that govern languages to comprehend how words and sentences are constructed. In NLP, these linguistic rules are transformed into patterns that machines can learn.

© Jonas Bjerg 2024
J. Bjerg, *The Early-Career Professional's Guide to Generative AI*,
https://doi.org/10.1007/979-8-8688-0456-4_3

One fundamental form of a language model is the unigram model, which treats each word in a text as a unit. It calculates the probability of each word appearing in the text without considering context or word order. Although this model is simple and computationally efficient, it fails to capture the intricacies and nuances of language.

Advanced bigram and trigram models were developed to overcome this limitation. These models consider pairs or triples of words, respectively, allowing them to capture a level of information and word order.

For instance, if we take the sentence "I love ice cream," a bigram model would analyze pairs, like "I love," "love ice," and "ice cream." On the other hand, a trigram model would consider triples such as "I love ice" and "love ice cream." This approach helps the model predict the likelihood of a specific word following a sequence of words.

However, it's important to note that bigram and trigram models have limitations. They can only capture an amount of context. Require assistance when dealing with long-range dependencies between words. Advanced language models have been developed to address these challenges, including neural networks (RNNs), long short-term memory networks (LSTMs), and the transformer architecture, which serves as the foundation for models like GPT-4.

These sophisticated models excel at understanding context over sequences of words, thereby generating text that closely resembles language. They are trained on amounts of text data to learn patterns and structures within language. Consequently, they can accurately predict the word in a sentence based on the preceding context.

The Mechanics of Language Models

To gain an understanding of language models, it is important to delve into their workings. Language models rely on concepts and processes, including tokenization, embeddings, and the architecture of the model.

Tokenization

Tokenization plays a role in training a language model. It involves breaking down the input text into more manageable parts known as "tokens."

In English, tokens can range from characters to words. The decision on size is significant as it influences how well the model performs and its ability to recognize patterns.

Now, let's explore further the implications of sizes. For instance, consider tokenizing at a character level.

When a model is divided into tokens at this level, it has the ability to generate text that closely resembles the training data in terms of spelling and grammar accuracy. However there is a drawback when the model tries to understand the meaning of word sequences; it can struggle in this aspect.

On the other hand, word-level tokenization presents a set of advantages and challenges. With this approach, the model is better equipped to grasp the nuances of meaning, how words are used in context. However, there is a trade-off. The model becomes more complex due to an increase in the number of tokens it needs to handle. This can result in complexity and size.

Embeddings

Once tokenization is completed, which involves breaking down text into fragments or tokens, these tokens need to be represented in a format that allows processing. This is where embeddings come into play. What exactly are embeddings? They are representations that not only convert tokens into numbers but also capture the underlying semantic meaning of those tokens.

By utilizing this approach, embeddings open up possibilities for the model to perform operations on text data, bridging the gap between language and machine computation.

Now, let's delve into a technique called Word2Vec, which is used to generate these embeddings. However, Word2Vec goes beyond word-to-vector conversion. It is a process that relies on the power of networks. The underlying idea behind Word2Vec is quite intuitive; it trains a network to predict a word based on its surrounding context or vice versa, deducing the context given a word. This prediction mechanism leads to the creation of vectors that accurately capture the semantic essence of words.

Essentially, Word2Vec surpasses representations and delves into capturing the intricate nuances and relationships between words. Through this approach, words with contexts or meanings are positioned closely together in the embedding space, enabling nuanced and context-aware computations by the model.

Model Architecture

The structure of the language model determines how it utilizes these embeddings to make predictions.

One common type of architecture is known as the recurrent neural network (RNN), which processes text in a certain manner. It maintains a state that captures information from all analyzed tokens.

However, RNNs face a limitation; they struggle to handle long-range dependencies due to the "vanishing gradient" problem. This means that the influence of information diminishes over time. To address this issue, long short-term memory (LSTM) networks were developed. LSTMs have a mechanism that allows them to retain information while disregarding details, making them more effective at processing lengthy text sequences.

Recently, the Transformer architecture has become the standard for language models. Unlike RNNs and LSTMs, Transformers process all tokens in the text simultaneously, making them more efficient. They also incorporate attention mechanisms, which enable the model to weigh the importance of each token when generating predictions.

GPT-4 utilizes a variant of the Transformer architecture known as the Transformer Decoder. This particular architecture processes text from left to right. It is well suited for tasks involving text generation.

In the following sections, we will explore networks and deep learning techniques in greater detail, providing a comprehensive understanding of how language models operate.

The History and Evolution of Language Models

The history of AI language models can be traced to Alan Turing's test for determining if a computer is capable of displaying intelligent behavior that is comparable to or indistinguishable from that of a human, which was developed in the early 1950s. The first successful application of NLP was ELIZA, which was developed by Joseph Weizenbaum in 1966. ELIZA was designed to mimic a psychotherapist by using simple pattern-matching rules.

In the following decades, statistical methods were introduced to improve NLP, including using hidden Markov models and probabilistic context-free grammar. However, it wasn't until the development of neural networks in the 1980s that significant progress was made in AI language models.

Today, AI language models are widely used for various applications such as chatbots, translation services, and content generation. The field of language modeling has made progress over time. We have gone from rule-based systems to the deep learning models we use today, each step getting us closer to creating machines that can understand and generate text that resembles human language. In this section, we will explore this evolution. Discuss some of the breakthroughs.

Rule-Based Systems

Rule-based systems evaluate text using specified rules and provide outputs or responses in accordance with predetermined parameters. For straightforward jobs like responding to oft-requested inquiries or giving basic details about a good or service, these systems perform admirably. However, rule-based systems have limitations when understanding complex natural language and generating creative responses.

During the stages of language model development, rule-based systems took the stage. What exactly are rule-based systems? These are models that rely on a predefined set of rules to perform tasks rather than learning from data. When it comes to language modeling, these rules mainly focus on structures and comprehensive vocabulary lists.

These early systems were proficient at producing text that adhered to norms. However, they had a limitation; they lacked the finesse and adaptability needed to comprehend and adjust to contexts. This meant that while the generated text might be technically accurate, it often lacked the smoothness and natural flow characteristic of speech and writing.

Let's take Google Translate as an example that many of us can relate to. When it was first introduced, users noticed that while the translations provided were acceptable, they often lacked the nuances and natural flow of the target language. These translations felt like they required intervention to refine them and make them sound more like something a human would say.

Google was well aware of this challenge. Sought to revolutionize their translation tool. Their goal was clear, not accuracy, but contextual relevance and natural-sounding output. To achieve this, Google's team embarked on a mission that led to the introduction of transformer models.

This wasn't an update; it represented a shift in language modeling. The inaugural paper on transformer models, which laid the foundation for language models, was a result of their dedicated efforts. Their aim was to surpass the limitations of rule-based systems and open doors to an era where models could understand and generate text that's both semantically accurate and contextually rich.

Statistical Language Models

Statistical models rely on probabilities derived from training data to make predictions. They use techniques such as n-grams, which analyze sequences of words, and hidden Markov models, which model the probability distribution over a sequence of observations. They are more flexible than rule-based systems and can generate more complex responses. However, they require large training data and may need to improve on unseen data.

Statistical language models emerged as a milestone in the evolving development of language models. Then, relying on rule-based systems, statistical language models utilized probability and statistics to predict words in a sequence.

These models relied on predicting the word in a sentence by considering the context provided by preceding words. Among models, n-gram models became prominent due to their widespread use. The term "n-gram" may sound technical. The concept is relatively straightforward. In an n-gram model, the sequence's previous "n" words influence predictions for the word. For example, a bigram (2-gram) model looks at one word for predictions, while a trigram (3-gram) model considers the last two words, and so on.

Statistical language models undoubtedly improved language prediction quality, with their foundations surpassing their rule-based predecessors. However, they still faced challenges in accounting for long-range dependencies.

Considering the limitations of n-gram models, which were confined to a fixed window of words (determined by "n"), they struggled with comprehending context from words back in the sentence. Consequently, they often needed to catch up on context that could be words behind, resulting in less accurate predictions.

Moreover, these models faced an issue known as the out-of-vocabulary (OOV) problem. In terms of encountering words that were not present in their training data, statistical language models found themselves at a loss. Being outside the model's vocabulary, these unfamiliar words posed a challenge. They frequently led to incorrect or suboptimal predictions.

In essence, while statistical language models marked an advancement in linguistics, they also emphasized the necessity for even more sophisticated and adaptable models to tackle the inherent challenges of language processing.

Neural Networks and Deep Learning

The strength of language models stems from their utilization of networks and deep learning techniques. These methods empower models to grasp patterns and structures within the data, enabling them to generate text that resembles expression.

In this section, we will give you an overview of these ideas and how they are utilized in language models.

Neural Networks

Neural networks are a type of machine learning model that takes inspiration from the brain. They consist of layers of nodes also known as "neurons" each performing a calculation on the data. The output from one layer of neurons becomes the input for the next layer enabling the network to transform data in countless ways.

The initial layer of a network is called the input layer, which receives the data (in the case of language models it would be embeddings). The final layer is known as the output layer for generating the predictions (in language models probabilities for the next word). Sandwiched between these two layers are one or more layers that carry out calculations.

To train networks, a process called backpropagation is employed. It involves adjusting connection weights between neurons to minimize discrepancies between the network's predictions and actual values. This enables learning patterns within data and improving predictions over time.

One of the main drawbacks of modern day neural networks is that while it is inspired by the neurons in our own brains, we haven't yet been able to store the learned information in the actual neurons. All information is currently being stored in the weights applied to the neuron between each layer. This is why generative AI companies are so protective of their model structure. If one managed to copy their weights, they would have close to a complete replica of the entire model. In an ideal world, the information should be stored in the neuron itself, that would enable the models to learn on data sets that are much much smaller. Similar to how you don't need to see thousands of hours learning how to ride a bike—like our current computer models do.

Deep Learning

Deep learning is an area within machine learning that focuses on networks with hidden layers, also known as deep neural networks. These networks have the ability to learn patterns surpassing their shallow counterparts. This makes them highly effective for tasks such as recognizing images, understanding speech, and processing language.

Regarding language models, deep learning enables the model to comprehend the structure of language. For example, individual words can be combined to form phrases, which in turn can be connected to create sentences and so forth. By leveraging networks, these hierarchical patterns can be learned and utilized for generating text that closely resembles human language.

In the following section, we will delve into applications where advanced language models are employed. We will discuss their usage in machine translation, sentiment analysis, text generation, and more. Additionally, we will provide examples that demonstrate how these applications work in practice.

Applications of Language Models

Language models have many applications, many of which we interact with daily, often without realizing. These applications span various industries and fields, demonstrating the versatility and potential of these models. In this section, we'll explore some of the key applications of language models.

Machine Translation

Machine translation is one of the earliest and most well-known language model applications. Services like Google Translate use language models to translate text from one language to another. These models are trained on a large bilingual text corpus and learn to map sentences in the source language to sentences in the target language. Modern machine translation systems, like Google's Neural Machine Translation, use advanced models like Transformers to handle complex translations and capture nuances in the text.

Sentiment Analysis

Sentiment analysis, often synonymous with opinion mining, is a fascinating application of language models that delves into emotions and opinions. At its core, sentiment analysis seeks to discern a text's underlying sentiment or emotional tone. But why is this important? In today's data-driven world, gauging the sentiment behind textual content offers profound insights, especially in sectors like marketing and customer service.

Understanding customer sentiment is equivalent to tapping into a goldmine of business feedback. Imagine a marketing team gauging the reception of a new product launch or a customer service department sifting through feedback to identify areas of improvement. With sentiment analysis, such tasks become streamlined. Businesses can quickly gauge their audience's general mood and sentiment by classifying textual feedback as positive, negative, or neutral.

However, the world of sentiment is intricate, and emotions aren't always black and white. Recognizing this, advanced language models have evolved to discern basic sentiments and delve into subtler emotional undertones. These state-of-the-art models are adept at identifying a spectrum of emotions, from joy and surprise to anger and sadness. Even more impressive is their capability to detect sarcasm—a complex expression that often juxtaposes positive words with a negative sentiment or vice versa.

Consider a statement like "Oh, great! Another flat tire." While "great" typically has a positive connotation, it's dripping with sarcasm in this context. Advanced language models trained for sentiment analysis can pick up on such nuances, ensuring that sentiments are analyzed in their true context.

In essence, sentiment analysis, powered by sophisticated language models, offers a lens into the emotional landscape of textual content. Whether for businesses aiming to refine their strategies or researchers aiming to understand human emotions better, sentiment analysis is a testament to the power of marrying linguistics with machine learning.

Text Generation

Diving into text generation uncovers an enthralling facet of language models, where machines don't just understand human language but actively craft it. Text generation, as the name suggests, revolves around the automated creation of textual content that mirrors human-like language patterns and nuances.

The spectrum of applications for text generation is vast and diverse. On one end, we have pragmatic tools designed to assist and enhance our daily written communications. Consider writing assistants like Grammarly. These aren't just about correcting grammatical errors; they harness the power of language models to suggest better phrasings, enrich vocabulary, and even ensure tone consistency. Similarly, smartphones' ubiquitous autocorrect and text prediction features are underpinned by language models that anticipate and suggest the next word or phrase, streamlining the typing process.

Venturing beyond the practical, text generation dives into the realm of creativity. With advanced language models at the helm, we witness the automated generation of artistic content like poetry, short stories, and even intricate scripts for plays or movies. These aren't just random assemblages of words; they're coherent narratives, often indistinguishable from human-written content.

A noteworthy example in this space is GPT-4, a cutting-edge language model developed by OpenAI. GPT-4 stands out not just for its sheer size but for its unparalleled prowess in text generation. It's not limited to producing short snippets; GPT-4 can craft extensive passages of text that maintain coherence, context relevance, and thematic consistency. Whether continuing a story from a given prompt, answering questions, or even generating technical content, GPT-4 showcases the zenith of what language models can achieve in text generation.

In summary, text generation, fueled by advanced language models, blurs the lines between human and machine-generated content. From practical writing aids to creative masterpieces, this application exemplifies the boundless potential of melding linguistics with artificial intelligence.

Information Extraction

Delving into vast oceans of textual data, information extraction is a beacon, helping discern the specific nuggets of information that matter most. Information extraction is sifting through textual content to pinpoint and extract precise data. This process has been significantly enhanced with the advent of sophisticated language models, evolving from a manual, labor-intensive task to an automated, efficient endeavor.

One of the foundational aspects of information extraction is entity recognition. In this context, entities refer to specific, identifiable elements within the text. This can encompass various elements, from names of individuals and geographical locations to dates, organizations, monetary values, and more. For instance, in a news article discussing a merger between two companies on a specific date, a language model can identify the names of the companies as entities and the date of the merger.

But information extraction goes beyond just recognizing entities. It delves deeper, aiming to understand the relationships and interplay between these entities. For instance, in a medical report that mentions a patient being prescribed a specific medication by a particular doctor, the model can discern the patient, medication, and doctor as entities and the relationship between them—that the doctor prescribed the medication to the patient.

The applications and implications of information extraction are profound, especially in sectors where time is of the essence and vast amounts of textual data need rapid analysis. Consider the legal domain. Legal professionals often grapple with voluminous case files, contracts, and legal documents. Information extraction can be a game-changer, allowing them to swiftly pinpoint key information, such as involved parties, dates, clauses, or precedents. Similarly, in the medical field, doctors and healthcare professionals can leverage this technique to quickly extract vital data from patient records, research papers, or clinical notes, aiding in diagnosis, treatment planning, and research.

In essence, information extraction, powered by advanced language models, represents a paradigm shift in how we approach and understand textual data. It's not just about reading and comprehending; it's about zooming in on the most crucial information, facilitating faster decision-making and deeper insights across various domains.

Chatbots and Virtual Assistants

In the digital age, where interactions often transcend the physical realm, chatbots and virtual assistants have emerged as the vanguard of human–computer communication. These digital entities, which have become household names like Siri, Alexa, and Google Assistant, are underpinned by powerful language models. But what exactly do these models contribute to our virtual conversationalists?

At their heart, chatbots and virtual assistants strive to simulate human-like interactions. When a user poses a question or issues a command, the language model deciphers the query, understands its context, and crafts an apt response. It's not just about retrieving information or executing commands; it's about doing so in a natural, intuitive, and conversational manner.

Imagine asking Siri about the weather, requesting Alexa to play your favorite song, or seeking directions from Google Assistant. The fluidity and relevance of their responses aren't mere coincidences; they result from intricate language models working behind the scenes, processing inputs, and generating outputs that align with human expectations.

However, it's imperative to recognize that chatbots and virtual assistants are just the tip of the iceberg when it comes to the applications of language models. With their ever-evolving capabilities, these models find utility in many sectors, from sentiment analysis and text generation to information extraction.

As advancements continue to reshape the landscape of language models, the horizon is rife with possibilities. We are on the cusp of witnessing these models harnessed in unprecedented ways, ushering in innovations that could redefine how we interact with technology and information.

Yet, like all technologies, language models have challenges and limitations. While their potential is undeniable, it's equally essential to approach them with a discerning lens, understanding their constraints and ensuring that their deployment is both responsible and beneficial. In the ensuing section, we will delve deeper into these challenges, offering a comprehensive perspective on language modeling.

Examples of Language models

In this section, we'll look at some examples of language models that programmers have recently developed. These models have a great deal of potential for the artificial intelligence community. These will probably keep changing and affect how we use technology.

BERT Language Model

Bidirectional encoder representations from transformers are abbreviated as BERT. Google created this transformer-based method for natural language processing. Neural language modeling techniques, such as transformer models, allocate attention to individual segments of an input. After that, the model determines which components of the input are most useful for interpreting the context and meaning. Specifically, the BERT language model is intended to train next-sentence prediction and language modeling in natural language processing software.

Autoregressive Language Model

A form of statistical modeling known as an autoregressive language model makes use of language input to predict the next word in a sequence. When determining which word should come before or after another, the model considers the context of just one word in the phrase. This modeling approach will take the forward or backward context into account. It can determine which word makes the most sense in between by looking at the word that comes before or after it. It will only take into account one context at a time, though. Consequently, this model is unable to decide solely on the entirety of a sentence.

Masked Language Model

Other language models can be trained to do NLP tasks using a masked language model. This is achieved by requesting the language model to correctly fill in the blanks of a sentence or phrase after concealing a portion of its words. To offer more context, these models assign keywords within a phrase more weight or attention than others. Alternatively, they might equally weigh each element to test the model's ability to provide a suitable phrase without providing any "hints."

PaLm Language Model

Google is working on a neural language model called the Pathways Language Model, or PaLM Language Model. Instead of being trained for a single goal, these 540 billion-parameter transformer models are being trained to do a variety of NLP-related tasks. This large-scale language model is expected to facilitate the scalability of language processing capabilities across various computers and technologies.

Acoustic Language Model

An acoustic language model is a type of neural language model that generates a set of plausible phonemes based on audio input. Similar to letters, phonemes are written characters that are meant to represent particular sounds. Multiple alphabetical letters can make up a phoneme (like the sound "CH"). The model can create written words from the combined phonemes using a pronunciation dictionary. A different type of language model then examines the phonemes to determine the most likely intended word sequence. A typical feature of automated speech recognition systems is this model. A variety of technologies, such as voice command capabilities and smart assistants use this approach.

Limitations and Challenges of Language Models

While language models have made significant strides and have many applications, they are not without their limitations and challenges. Understanding these is crucial for developing and using these models responsibly. This section will discuss some of the fundamental limitations and challenges associated with language models.

Understanding vs. Generating Text

One of the key limitations of current language models reveals a fascinating yet crucial distinction: the difference between generating text and truly understanding it. Modern language models, equipped with vast amounts of data and sophisticated architectures, excel at crafting text that, on the surface, mirrors human language in its complexity and nuance. However, this prowess in text generation only sometimes translates to genuine comprehension, and this distinction is pivotal.

At the heart of human understanding lies a rich tapestry of experiences, emotions, knowledge, and context. When humans read or hear a text, they don't just process words; they relate them to a broader understanding of the world, drawing from personal experiences, cultural contexts, and a deeply ingrained sense of causality and logic.

Contrast this with language models. Despite their computational sophistication, they lack a genuine "worldview." They don't possess an innate understanding of the intricacies of the world, the cause-and-effect relationships that govern events, or the common sense that humans acquire over a lifetime of experiences. Instead, their knowledge reflects the vast amounts of data they've been trained on. They recognize and generate text based on patterns and structures they've gleaned from this data rather than any innate understanding.

37

This distinction manifests in multiple ways. While a language model can produce text that sounds impressively human-like, it can just as easily churn out content that, upon closer inspection, is nonsensical or devoid of factual accuracy. For instance, a model might craft a grammatically flawless sentence about a historical event that never occurred or propose a solution to a problem that defies basic logic.

The chasm between understanding and generating text underscores a vital aspect of language models: they are tools and reflections of the data they've been trained on. While they can emulate human-like language generation, true comprehension—rooted in experience, emotion, and a deep-seated understanding of the world—remains a uniquely human domain. As we continue to develop and deploy these models, we must recognize their capabilities and limitations, ensuring that we harness their strengths while remaining cognizant of their constraints.

Bias in AI

In the digital era, as artificial intelligence seeps into the fabric of everyday life, the challenge of bias in AI emerges as a paramount concern. The essence of AI, particularly in language models, lies in its ability to learn from vast troves of data. But herein lies a critical caveat: what if the data carries inherent biases?

Language models, by virtue of their design, are mirrors. They reflect the nature, nuances, and, unfortunately, the prejudices of the data they are trained on. When this data, often sourced from the vast expanses of the Internet, contains biases—be they subtle or overt—these models can inadvertently internalize and perpetuate them. The result? AI outputs that, at times, echo sentiments that are sexist, racist, or discriminatory in myriad other ways.

This isn't just a theoretical concern. There have been tangible instances where AI models, left unchecked, have produced outputs that betray deep-seated biases, thereby perpetuating harmful stereotypes or offering skewed perspectives.

Recognizing the gravity of this challenge, AI ethics has taken center stage in recent years. Researchers, ethicists, and technologists are joining hands, aiming to untangle the web of biases that might lurk in AI models. Their objectives are multifold:

- **Detection:** The first step to addressing bias is recognizing its presence. Tools and methodologies are being developed to scrutinize AI models, identifying instances where they might produce biased or discriminatory outputs.

- **Mitigation:** Once biases are detected, efforts shift toward mitigation. This could involve refining the training data, retraining the model, or implementing algorithms that actively counteract biases.

- **Awareness and Education:** Beyond the technical aspects, there's a growing emphasis on raising awareness about AI bias. By educating AI developers, users, and the broader public, the goal is to foster a collective responsibility toward creating AI systems that are fair and equitable.

In conclusion, while AI offers unparalleled opportunities with its data-driven prowess, it also presents challenges that are deeply intertwined with the fabric of society. Bias in AI reflects historical and societal biases, and addressing it requires a concerted, multidimensional approach. As we stand on the cusp of an AI-driven future, ensuring that these systems are unbiased and equitable becomes a technical challenge and a moral imperative.

Data Privacy and Security

In a world increasingly governed by data, the twin pillars of data privacy and security have taken on paramount importance, especially in the realm of artificial intelligence and language models.

The very foundation of language models lies in their voracious appetite for data. They are fed vast quantities of text data, enabling them to learn, predict, and generate human-like text. While this offers immense potential in terms of capabilities, it also introduces significant vulnerabilities.

At the forefront of these concerns is the sanctity of personal data. While training datasets are usually processed to be anonymous, stripping away personally identifiable information (PII), the sheer depth and breadth of these datasets can sometimes lead to unforeseen challenges. There exists the very real risk that a sophisticated language model, when queried in a particular way, might inadvertently generate outputs that echo sensitive information present in its training data. Such outputs, even if unintentional, could have serious implications for individual privacy.

Moreover, in applications where language models interact directly with user-generated content—think chatbots, virtual assistants, or personalized recommendation systems—the stakes are even higher. These systems could potentially be exposed to a myriad of personal details, from benign preferences to critical personal data. If not properly secured and managed, this data could be vulnerable to breaches, misuse, or unauthorized access.

Addressing these challenges requires a multipronged approach:

- **Robust Anonymization:** While datasets are typically anonymized, there's a need for even more robust methods that ensure no trace of personal data can be reconstructed or inferred.

- **Secure Storage and Processing:** Ensuring that data, especially user-generated content, is stored and processed securely is crucial. This involves encrypted storage solutions, secure data transmission protocols, and regular security audits.

- **Transparency and Consent:** Users need to be informed about how their data will be used and should have the agency to provide (or withhold) consent. Transparent data policies can foster trust and ensure compliance with data protection regulations.

- **Continuous Monitoring:** Even post-deployment, language models should be continuously monitored to detect and rectify any inadvertent generation of sensitive information.

In essence, as the capabilities of language models soar, so does the responsibility of ensuring they operate within the bounds of data privacy and security. Balancing the promise of AI with the imperatives of data protection is a challenge that will define the trajectory of AI ethics in the years to come.

Resource Intensity

The marvels of modern language models, with their unparalleled capabilities and sophistication, come at a cost, and it's not just metaphorical. The journey from conceptualization to deployment of these models entails an immense consumption of computational resources, bringing to the forefront concerns that are both environmental and socioeconomic in nature.

- **Environmental Implications:** Training state-of-the-art language models demands vast computational power. These computations are typically performed on powerful GPUs or TPUs housed in data centers. While these centers are marvels of modern engineering, they are also voracious consumers of electricity. Given the prolonged training times for sophisticated models, the energy consumption is substantial. This, in turn, has a direct environmental impact, especially if the energy sources are non-renewable. The carbon footprint of training a single large model can be equivalent to the emissions from multiple cars over their entire lifetimes!

- **The concentration of Power:** The resource intensity of training large language models also has broader societal implications. The computational infrastructure required is expensive, both in terms of hardware costs and energy bills. This economic barrier means that the ability to develop and refine cutting-edge models is often restricted to well-funded organizations, be they tech giants or affluent research institutions. This concentration of capability risks leading to a concentration of influence and power. If only a few entities can shape the evolution and deployment of AI, it raises questions about equitable access, representation, and influence in the AI-driven future.

- **Democratization Concerns:** The economic barriers also impede the democratization of AI research. Independent researchers, startups, or institutions in economically disadvantaged regions might find it challenging to contribute to or benefit from advancements in language models, potentially widening the technological divide.

In conclusion, while the achievements of large language models are undeniably impressive, the resource intensity associated with their development serves as a reminder of the broader implications. Balancing the pursuit of AI excellence with environmental responsibility and equitable access is a challenge that the AI community, and society at large, must grapple with in the coming years.

Misuse of Technology

In the pantheon of technological advancements, every innovation carries a dual potential: the promise of progress and the peril of misuse. Language models, with their impressive capabilities, are no exception. Their prowess in generating human-like text, while transformative in many positive ways, also opens the door to a spectrum of malicious applications.

Potential Malfeasance with Language Models:

- **Fake News Generation:** In an age where information is power, the potential for spreading misinformation or disinformation is a significant concern. Language models can craft news articles or reports that are entirely fictitious yet sound authentic, further exacerbating the challenges of distinguishing fact from fiction in the digital age.

- **Spam and Malware:** The age-old problem of spam emails gets a new twist with advanced language models. Instead of generic, easily detectable spam content, these models can generate personalized, contextually relevant spam that's harder to filter out.

- **Phishing Attacks:** Phishing emails thrive on deceit, trying to masquerade as legitimate communication to extract sensitive information. Language models can craft eerily convincing phishing emails tailored to individual targets, making them more challenging to detect and increasing the risk of successful scams.

Addressing the Challenge: The threat of misuse isn't one that can be countered with a singular solution. It demands a multifaceted approach:

- **Technical Solutions:** Advances in AI can be harnessed not just to generate content but also to detect and counter misuse. Misuse detection systems can be trained to identify content generated by language models, flagging potential spam, phishing attempts, or fake news articles. However, this often becomes a cat-and-mouse game, with detection systems and generation models continuously evolving to outdo each other.

- **Policy Solutions:** Beyond the realm of technology, there's a pressing need for robust policy frameworks. Regulations and guidelines can set boundaries on the deployment and use of language models, ensuring accountability and establishing deterrents against misuse.

- **Public Awareness:** Education and awareness play a crucial role. By educating the public about the capabilities (and potential misuses) of language models, individuals can be better equipped to scrutinize and discern genuine content from AI-generated deceit.

In summation, the advent of advanced language models heralds a new era of opportunities and challenges. As we harness their potential for progress, it's imperative to remain vigilant against the shadows of misuse, ensuring that the promise of AI is realized without compromising integrity, security, or trust.

In the next section, we'll look ahead to the future of language models. We'll discuss ongoing research in the field and the potential for advancements in areas like transfer learning, unsupervised learning, and more. Despite the challenges and limitations, the future of language models is promising, and the potential applications are vast.

The Future of Language Models

As we look to the future of language models, we see a landscape filled with potential and opportunities for further advancements. Despite the challenges and limitations we've discussed, ongoing research and development in the field promise to push the boundaries of what language models can achieve. This section will explore some areas where we might see significant advancements in the coming years.

Transfer Learning

In the vast and intricate landscape of machine learning and artificial intelligence, transfer learning emerges as a beacon of efficiency and adaptability. But what exactly does this technique entail, and why is it garnering such attention and acclaim?

- **Conceptualizing Transfer Learning:** Imagine training for a marathon. The stamina, strength, and endurance you develop don't just benefit your marathon running; they can be "transferred" to enhance your performance in, say, hiking or cycling. Transfer learning in AI operates on a similar principle. Instead of starting the learning process from scratch for every new task, models can capitalize on knowledge acquired from previous tasks, adapting and fine-tuning it for new challenges.

- **The Rise of BERT, GPT-4, and Beyond:** Models like BERT (Bidirectional Encoder Representations from Transformers) and GPT-4 (Generative Pre-trained Transformer 4th version) have thrust transfer learning into the spotlight. BERT, for instance, is pre-trained on vast amounts of text data to understand language structure and context. This pre-trained model can then be fine-tuned for specific tasks, such as sentiment analysis or question-answering, with relatively minimal additional training. GPT-4, another behemoth in the AI world, leverages its extensive pre-training to generate coherent and contextually relevant text across a plethora of prompts and tasks.

Advantages and Future Prospects:

- **Efficiency:** Transfer learning reduces the need for extensive training data for specific tasks. By leveraging knowledge from prior training, models can achieve competitive, if not superior, performance with significantly less data.

- **Computational Savings:** Training large models from scratch is resource-intensive, both in terms of time and computational power. Transfer learning offers a shortcut, necessitating training only for the fine-tuning phase, thereby conserving resources.

- **Versatility:** Transfer learning paves the way for models that are not just efficient but also versatile, capable of adapting to a myriad of tasks with minimal retraining.

With the successes of models like BERT and GPT-4 as a testament, the future of transfer learning looks promising. As research in this domain accelerates, we can anticipate models that are even more efficient, adaptable, and capable, pushing the boundaries of what AI can achieve across diverse tasks and challenges.

Unsupervised Learning

Unsupervised learning is a type of machine learning where a model learns to identify patterns in data without any labeled examples. Unsupervised learning is particularly relevant for language models, as much of the text data available for training is unlabeled. Advances in unsupervised learning could lead to more powerful language models that can understand and generate text more effectively.

The realm of machine learning is vast and varied, with different techniques tailored to different challenges. Among these, unsupervised learning stands out as a particularly intriguing and potent approach, especially when navigating the nuances of language.

- **Diving into Unsupervised Learning:** At its core, unsupervised learning is akin to exploration without a map. Traditional supervised learning relies on labeled data, where each input is paired with a corresponding output, guiding the model toward the desired answer. Unsupervised learning, in contrast, ventures into the data wilderness without such labels, seeking to uncover hidden patterns, structures, or relationships on its own.

- **The Relevance for Language Models:** Language, in all its richness, is a vast ocean of data. While some of this data comes with annotations or labels, a significant portion is unlabeled. Think of the billions of words on the Internet—articles, blogs, social media posts, and more. Unsupervised learning is tailor-made for such scenarios, allowing language models to immerse themselves in this data, seeking to understand the intricacies of language without explicit guidance.

Potential Outcomes and Advances:

- **Clustering and Topic Modeling:** One of the primary applications of unsupervised learning in language is clustering or topic modeling. For instance, given a vast collection of articles, an unsupervised model might group them based on themes or subjects, even if these articles weren't categorized beforehand.

- **Word Embeddings:** Techniques like Word2Vec, though not entirely unsupervised, leverage vast unlabeled text data to create dense vector representations of words, capturing their semantic meanings.

- **Dimensionality Reduction:** Unsupervised models can identify the most critical features or dimensions in data, enabling more efficient processing and analysis.

- **Generating Text:** As unsupervised models become more sophisticated, their ability to generate coherent and contextually relevant text can rival, if not surpass, models trained with supervised techniques.

With the burgeoning amount of text data and the limited availability of labeled datasets, unsupervised learning is poised to play an increasingly pivotal role in the evolution of language models. As research in this domain intensifies, we can look forward to models that not only understand the nuances of language more deeply but can also craft text with greater finesse and authenticity.

Fine-Tuning and Personalization

Fine-tuning is a process where a pre-trained model is further trained on a specific task to improve its performance. This process can also be used to personalize a language model for a specific user, allowing it to understand better and generate text in the user's style. As we move toward more personalized AI, we can expect to see advancements in fine-tuning techniques.

Fine-tuning and personalization have emerged as essential facets in the quest for more effective and user-centric models. But what do these processes entail, and why are they so pivotal in the current AI landscape?

- **A Deeper Dive into Fine-Tuning:** Imagine purchasing a tailor-made suit. While it might fit reasonably well off the rack, a few adjustments can make it fit perfectly. Similarly, in the world of AI, fine-tuning operates on this principle. While a model pre-trained on vast amounts of data has a broad understanding, fine-tuning it on specific data related to a particular task refines its capabilities, allowing it to perform that task with enhanced precision. It's like customizing the model to fit the "shape" of a specific problem or dataset.

- **The Magic of Personalization:** Beyond task-specific adjustments, fine-tuning can also be harnessed for a more personal touch: tailoring models to individual users. Here's where the magic of personalization comes into play. By fine-tuning a language model on a user's writings, communications, or preferences, the model can start to "speak" and "understand" in a style that's uniquely tailored to that user. Whether it's capturing a user's tone, vocabulary preferences, or even idiosyncratic expressions, personalized models can offer a more intuitive and seamless interaction experience.

The Road Ahead:

- **Adaptive Learning:** As AI systems become more sophisticated, we might see models that can fine-tune themselves in real time, adapting to a user's changing preferences and needs.

- **Ethical Considerations:** With personalization comes the challenge of data privacy. Ensuring that user data used for fine-tuning is handled securely and ethically will be paramount.

- **Diverse Applications:** From personalized content recommendations to tailored virtual assistants and bespoke writing aids, the applications of fine-tuned models are vast and varied.

In conclusion, as the narrative of AI shifts toward offering more personalized and user-centric experiences, fine-tuning and personalization will be at the heart of this transformation. By honing models to cater to specific tasks and individual users, we're not just enhancing AI's capabilities; we're making it more intuitive, relatable, and user-friendly, bridging the gap between machine intelligence and human uniqueness.

Ethics and Fairness

As we've discussed, bias in AI is a significant challenge. In the future, we expect to see more research and development focused on making language models more ethical and fairer. This could involve techniques for detecting and mitigating bias in the training data, as well as methods for ensuring that the model outputs are fair and unbiased.

While AI's prowess and potential are undeniable, ensuring that its capabilities are harnessed in a just and equitable manner is paramount. Let's delve deeper into the challenges and prospective solutions in this domain.

The Ethical Imperative: AI models, particularly language models, derive their knowledge from vast amounts of data. However, this data, often sourced from human-generated content, may carry with it the biases, prejudices, and disparities of the society it originates from. If unchecked, these biases can be internalized and perpetuated by the AI systems, leading to outputs that may be discriminatory or unjust.

The Path to Fairness: Achieving fairness in AI is a multifaceted endeavor, encompassing various strategies and initiatives:

- **Bias Detection:** The first step toward a fair AI system is recognizing the presence of biases. Tools and algorithms are being developed to scrutinize both the training data and the AI model's outputs, flagging potential biases or disparities.

- **Data Rectification:** Once biases are identified in the training data, efforts can focus on rectifying them. This might involve enriching the dataset with underrepresented data, rebalancing skewed distributions, or even synthetically generating data to ensure a more balanced representation.

- **Model Regularization:** Techniques can be employed during the training process to penalize and diminish biased predictions. Regularization methods can ensure that models don't overly rely on potentially biased features or patterns.

- **Post hoc Analysis:** Even after a model has been trained, its outputs can be analyzed and adjusted to ensure fairness. Post-processing techniques can recalibrate predictions to ensure they align with fairness criteria.

- **Transparency and Interpretability:** Ensuring that AI models are transparent and interpretable can help stakeholders understand how decisions are made, shedding light on potential biases and providing avenues for rectification.

- **Stakeholder Involvement:** Engaging diverse stakeholders in the development and evaluation process can offer varied perspectives, ensuring that fairness considerations are holistic and representative.

Looking Forward: The future of AI ethics and fairness is rife with challenges but also with opportunities. As research intensifies, we can anticipate more sophisticated techniques for ensuring fairness, coupled with industry standards and best practices. Additionally, regulatory frameworks might emerge, guiding and overseeing the development of ethical AI systems.

In essence, as we stand on the cusp of an AI-driven era, the imperatives of ethics and fairness cannot be overstated. Ensuring that AI systems are just, equitable, and unbiased is not just a technical challenge; it's a societal responsibility, defining the contours of the AI landscape for generations to come.

Regulation and Policy

Finally, as language models become more prevalent and impactful, we expect to see more focus on regulation and policy. This could involve guidelines for the responsible use of language models, regulations to prevent misuse, and policies to ensure data privacy and security.

The meteoric rise of language models in various domains of our digital life underscores not just their potential but also the need for oversight. As with any powerful technology, the deployment and utilization of language models carry inherent risks and responsibilities. Hence, the establishment of robust regulations and policies becomes essential to guide this burgeoning field toward a future that's both innovative and secure.

The Imperative for Oversight:

- **Responsible Use:** As language models influence everything from content recommendations to news generation, guidelines are needed to ensure their applications align with ethical and societal norms.

- **Prevention of Misuse:** The ability of language models to generate human-like text can be weaponized for malicious purposes, such as creating fake news or phishing emails. Regulations can act as deterrents and safeguards against such misuse.

- **Data Protection:** Given the vast amounts of data that fuel these models, policies ensuring data privacy, consent, and security are paramount to protecting individual rights and maintaining public trust.

Potential Avenues for Regulation and Policy:

- **Transparency Standards:** Regulatory bodies could mandate that developers and deployers of language models disclose the training data sources, methodologies, and objectives, ensuring clarity and accountability.

- **Auditing Mechanisms:** Regular audits of language models, especially those deployed in critical areas like health care or finance, can ensure they adhere to quality and fairness standards.

- **Redressal Mechanisms:** Policies can establish frameworks for users or stakeholders to report issues, biases, or potential misuse of language models, coupled with avenues for redressal or rectification.

- **International Collaboration:** Given the global nature of AI and its applications, international collaboration can harmonize regulations, ensuring consistent standards and preventing regulatory arbitrage.

- **Ethical Guidelines:** Beyond strict regulations, the development of ethical guidelines can guide researchers, developers, and businesses in the responsible creation and deployment of language models.

- **Data Rights:** Regulations can clarify data ownership, consent mechanisms, and rights to erasure or correction, ensuring users have agency over their data and its use.

The Road Ahead: While the potential of language models is vast, their responsible and ethical deployment is a collective responsibility that spans developers, businesses, regulators, and society at large. Crafting a balanced regulatory framework is a challenge—one that must navigate the tightrope between fostering innovation and ensuring safety, privacy, and fairness.

In the coming years, as the capabilities and influence of language models grow, the dialogue surrounding their regulation and policy will intensify. Through collaborative efforts, we can chart a course toward a future where language models are not just powerful tools but also pillars of trust, equity, and societal well-being.

Conclusion

The future of language models is promising. Despite the challenges and limitations, the potential applications and advancements are vast. As we continue to push the boundaries of what these models can achieve, we can expect them to play an increasingly important role in our lives.

CHAPTER 4

The Unexpected Evolution of AI

As we explore the unexpected evolution of AI, it's instructive to first look back at the predictions made by experts in the past. These predictions, while not consistently accurate, provide a fascinating glimpse into our evolving understanding of AI and its potential impact on society.

In the early days of AI, there was a common belief that the technology would first and foremost impact blue-collar jobs. These jobs typically involve manual labor, such as factory work, construction, and driving. The rationale behind this prediction was straightforward: many blue-collar jobs involve repetitive tasks that seemed well-suited for automation. Machines could perform these tasks more efficiently and tirelessly than humans, increasing business productivity and cost savings.

Following the automation of blue-collar jobs, experts predicted that AI would begin to impact low-scale white-collar jobs. These jobs involve non-manual labor and typically require a certain level of education or training. Examples include data entry clerks, customer service representatives, and administrative assistants. These jobs often involve routine tasks that, while not physical, could be automated using more advanced AI algorithms.

© Jonas Bjerg 2024
J. Bjerg, *The Early-Career Professional's Guide to Generative AI*,
https://doi.org/10.1007/979-8-8688-0456-4_4

Next in line, according to these predictions, were the high-IQ white-collar jobs. These jobs require expertise and intellectual work, such as doctors, lawyers, and engineers. While these jobs involve complex tasks that seem less susceptible to automation, AI could eventually take over certain aspects, such as diagnosing diseases or drafting legal documents.

Finally, experts predicted that the last jobs to be impacted by AI would be creative jobs. These jobs involve creating new ideas, artworks, or designs, such as artists, writers, and designers. The belief was that these jobs, which rely heavily on human creativity and intuition, would be the hardest to automate.

These predictions painted a clear picture of the expected evolution of AI: from manual labor to intellectual work and finally to creative endeavors. However, as we'll see in the next section, the reality of AI evolution has turned out to be quite different.

The Reality of AI Evolution

The predictions of the past, while logical in their progression, still needed to materialize as expected fully. The reality of AI evolution has been counterintuitive, with creative jobs among the first to be impacted by AI. This unexpected turn of events can be attributed to several factors, including the rapid advancement of AI capabilities and the unique characteristics of creative tasks.

Contrary to the belief that creative jobs would be the last frontier for AI, we've seen AI make significant inroads into creative fields. From AI-generated artwork and music to AI-written articles and scripts, the creative sector has seen a surprising level of automation. This is not to say that AI has replaced human creativity, but rather, it has become a tool that can mimic certain aspects of the creative process.

One of the reasons for this unexpected development is the nature of creative tasks. While creativity is often considered a uniquely human trait, many creative tasks involve patterns and structures that can be learned. For instance, a piece of music has a certain rhythm and melody, a painting has composition and color patterns, and a written article has grammatical rules and stylistic conventions. These patterns and structures can be learned by an AI, allowing it to generate creative outputs.

Another factor is the rapid advancement of AI capabilities, particularly in natural language processing. Language models like GPT-4 have shown an impressive ability to generate human-like text, opening up new possibilities for AI in creative writing. These models can write articles, generate ideas, draft emails, and even write code, tasks that were once thought to be the exclusive domain of humans.

The digital nature of many creative tasks has further amplified the impact of AI on creative jobs. Unlike physical tasks, which require the manipulation of physical objects, many creative tasks involve digital inputs and outputs. This makes them more accessible to AI, as they can be easily broken down into data that an AI can process.

In the next section, we'll delve deeper into the role of language models in this unexpected evolution of AI. We'll explore how these models work, why they're particularly suited to creative tasks, and what this means for the future of AI.

The Role of Language Models

Language models have played a pivotal role in the unexpected evolution of AI. These models, designed to predict the likelihood of a sequence of words, have proven remarkably effective at generating human-like text. This has opened up new possibilities for AI in creative fields, challenging our traditional notions of creativity and automation.

At their core, language models are statistical models that learn the patterns and structures of a language. They are trained on vast amounts of text data, learning the statistical patterns that underpin the language. This allows them to predict the next word in a sentence with high accuracy based on the context provided by the preceding words.

While this might seem like a simple task, it has profound implications for the capabilities of AI. A language model can generate coherent and contextually relevant sentences by accurately predicting the next word in a sentence. This allows it to write articles, generate ideas, draft emails, and even write code, tasks that were once thought to be the exclusive domain of humans.

The power of language models comes from their ability to learn and mimic human language patterns. Language is a creative endeavor. It involves combining words in novel ways to express ideas, tell stories, and convey emotions. By learning the patterns of language, a language model can mimic this creative process, generating text that closely resembles human-written text.

However, it's important to note that while language models can mimic the creative process, they need help understanding the text they generate. They do not have a concept of the world, do not understand causality, or have common sense. They generate text based on patterns they have learned from their training data. While this can often produce impressive results, it can also lead to nonsensical or factually incorrect outputs.

Despite these limitations, the capabilities of language models have had a significant impact on creative jobs. They have opened up new possibilities for automation, challenged our traditional notions of creativity, and raised important questions about the future of work. In the next section, we'll explore these implications further, providing case studies of industries where AI has had a significant impact.

Case Studies

To better understand the impact of AI on creative jobs, let's delve into some specific case studies. These examples, drawn from various industries, illustrate the transformative power of AI and the unexpected ways in which it has influenced the world of work.

Case Study 1: Journalism

The advent of artificial intelligence in the realm of journalism marks a fusion of cutting-edge technology with one of the oldest forms of mass communication. Newsrooms across the globe, from venerable institutions like the Associated Press to financial giants like Bloomberg, are harnessing the power of AI to craft news stories. But how exactly is this transformation taking shape, and what are its implications?

The AI Newsroom

- **Automating the Formulaic:** Certain news topics, such as financial earnings reports or sports scores, follow a structured and predictable format. AI excels at ingesting structured data and converting it into coherent narratives, making such topics prime candidates for automation. For instance, an AI system can quickly analyze a company's quarterly earnings data and generate a concise report, highlighting key figures and trends.

- **Speed and Volume:** In the fast-paced world of news, timeliness is crucial. AI-driven journalism tools can churn out articles within moments of receiving the requisite data. This rapidity, coupled with the ability to produce articles in bulk, offers news organizations a competitive edge, especially when covering real-time events or breaking news.

- **Cost Efficiency:** Automating certain segments of news production can lead to operational efficiencies and cost savings, allowing news agencies to allocate human resources to more complex, investigative, or nuanced stories.

Challenges and Considerations

- **Depth and Nuance:** While AI can craft accurate and structured reports, it lacks the human touch needed for in-depth analysis, emotional resonance, or capturing the subtleties of a story. Human journalists bring context, perspective, and a deeper understanding of sociopolitical nuances, elements that AI, in its current form, struggles to replicate.

- **Ethical Concerns:** The use of AI in journalism raises several ethical questions. How should news agencies disclose the use of AI to their readers? Is there a risk of bias in AI-generated content, especially if the underlying data or algorithms carry inherent biases?

- **Job Displacement:** There's an ongoing debate about the impact of AI on employment in journalism. While automation can lead to efficiencies, there's also concern about potential job losses or the devaluation of human-driven investigative journalism. It is a highly contentious issue. It is crucial for the industry to find a balance, where AI automates repetitive tasks and data-driven reporting, enabling journalists to focus on areas that require human insight and common sense.

Looking Ahead

As AI continues to make strides in journalism, a symbiotic relationship between human journalists and AI tools is emerging. Instead of viewing AI as a replacement, it's being seen as a complementary tool, automating repetitive tasks and freeing up human journalists to delve into deeper, more impactful stories.

In conclusion, the interplay between AI and journalism is a testament to how technology can reshape industries. By striking a balance between automation and the human touch, the future of journalism can be both technologically advanced and rich in depth and perspective.

Case Study 2: Music Composition

The incursion of AI into the realm of music composition heralds a new era of creative potential and accessibility. AI's role in music is expanding beyond mere experimental projects to become an integral component of the music industry, offering solutions for background scores, theme music, and soundtracks for videos, games, and various forms of digital content. Companies like Jukedeck, Amper Music, and others are at the forefront, harnessing AI to create music that adapts to specific needs with remarkable efficiency and creativity. This burgeoning field, while not intended to supplant human composers, provides valuable tools that democratize music composition and offer innovative ways to enhance the auditory experience of multimedia projects.

The Process and Potential

- **Analyzing and Generating Music:** At the heart of AI music composition is the technology's ability to analyze vast amounts of musical data, learning from existing compositions across genres and styles. AI algorithms, particularly those based on deep learning and neural networks, can identify patterns, structures, and elements that define different moods, atmospheres, and energy levels in music. This capability allows AI to generate new compositions that match specific criteria, such as tempo, key, mood, and even complexity, tailored to complement a video's narrative or a game's dynamics.

- **Bridging Gaps in Accessibility and Creativity:** One of the most significant advantages of AI in music composition is its accessibility. Independent filmmakers, game developers, and content creators, who may not have the resources to commission original music scores, find in AI a cost-effective solution to obtain custom music that aligns with their creative vision. Furthermore, AI-generated music can be quickly adapted and modified, providing creators with flexibility and control over the final product, ensuring the music perfectly fits the intended purpose.

The Collaborative Synergy Between AI and Human Musicians

- **Extending Creative Horizons:** While AI can produce music autonomously, its most transformative potential lies in collaboration with human musicians and composers. AI can offer novel musical ideas, motifs, and harmonies that can inspire musicians, pushing the boundaries of traditional composition and opening up new creative horizons. This symbiotic relationship allows for the exploration of uncharted musical landscapes, where AI-generated elements are woven into compositions to create complex, captivating pieces that might not have been conceivable by human minds alone.

- **Educational Tool and Creative Partner:** AI in music composition serves as an invaluable educational tool, offering students and aspiring composers the opportunity to experiment with music creation without

the need for extensive musical training. AI platforms
can provide instant feedback and suggestions,
making the learning process interactive and engaging.
Moreover, seasoned composers can use AI as a creative
partner, employing its capabilities to explore new
musical styles or to expedite the compositional process
by generating ideas that can be further refined and
developed.

Challenges and Ethical Considerations

- **Authenticity and Emotion:** One of the ongoing
  debates surrounding AI in music composition concerns
  the authenticity and emotional depth of AI-generated
  music. While AI can replicate patterns and styles, the
  nuanced expression and emotional intent inherent in
  music created by human composers can be elusive.
  Addressing this challenge involves enhancing AI's
  ability to interpret and convey emotions through music,
  a frontier that continues to evolve as AI technologies
  advance.

- **Intellectual Property and Authorship:** The
  rise of AI-generated music also raises questions
  about intellectual property rights and authorship.
  Determining the ownership of AI-created
  compositions, particularly when they are derived
  from analyzing existing works, is a complex issue
  that intersects with copyright law, ethics, and the
  evolving definitions of creativity and authorship in the
  digital age.

Looking Ahead

As AI technology continues to advance, its role in music composition is set to grow, fostering new forms of musical expression and collaboration between humans and machines. The future promises an enriched musical landscape where AI-generated compositions coexist with human-created music, each complementing the other and expanding the possibilities of what music can be. By embracing AI as a tool for creativity and innovation, the music industry can unlock unprecedented opportunities for composers, musicians, and creators, paving the way for a future where music transcends traditional boundaries and becomes more accessible and diverse than ever before.

In conclusion, AI's impact on music composition is profound, offering both opportunities and challenges that will shape the future of music. As the technology matures and its integration within the creative process deepens, AI will continue to redefine music composition, bringing forth a new era of artistic collaboration that leverages the best of human creativity and machine intelligence.

Case Study 3: Legal Services

The legal realm, with its intricate webs of statutes, precedents, and legalese, might seem an unlikely candidate for technological disruption. Yet, artificial intelligence is carving a niche for itself, reshaping how legal professionals operate and redefining efficiencies in this venerable field.

AI's Foray into Law

- **Streamlining Legal Research:** Legal research, a cornerstone of any legal proceeding, involves sifting through vast troves of statutes, case laws, and legal journals. AI-driven platforms, like ROSS Intelligence, harness natural language processing to understand queries and fetch relevant legal precedents or statutes with precision, dramatically reducing the time lawyers spend on research.

- **Contract Analysis:** Contracts, with their standardized formats but crucial nuances, are ripe for AI intervention. Solutions like LawGeex employ AI to scrutinize contracts, identify potential pitfalls or deviations from standard clauses, and even suggest edits. This not only ensures accuracy but also speeds up the contract review process.

- **Predictive Analysis:** Some AI tools delve into the realm of predictive analytics, analyzing past legal cases to predict potential outcomes of current cases and offering lawyers insights into how a case might unfold based on historical data.

Implications and Challenges

- **Increased Efficiency:** By automating tasks that are routine yet time-consuming, lawyers can focus on more strategic aspects of a case, such as formulating arguments or client counseling.

- **Cost Savings:** Automation can lead to cost savings for legal firms, which can potentially translate to reduced legal fees for clients.

- **Consistency and Accuracy:** AI-driven tools, with their ability to analyze vast amounts of data quickly, can bring a level of consistency and accuracy to tasks like contract review, minimizing human oversight.

- **Ethical and Reliability Concerns:** The deployment of AI in legal matters raises questions about its reliability and the ethical considerations of relying on algorithms for crucial decisions. Incorrect predictions or oversights by an AI tool could have serious legal repercussions.

- **Job Impact:** Similar to other fields, there's a debate about AI's impact on employment within the legal industry. While some routine tasks might be automated, the nuanced, strategic, and interpersonal aspects of legal work remain firmly in the human domain.

The Road Ahead

The intersection of AI and legal services is a testament to technology's pervasive influence. As AI tools become more sophisticated, their integration into legal workflows is set to deepen. However, a balanced approach, where AI augments human expertise rather than replaces it, is likely to dominate. In essence, while AI can sift through legal tomes or analyze contracts, the art of legal strategy, negotiation, and advocacy remains an intrinsically human endeavor.

Case Study 4: Graphic Design

The canvas of graphic design, traditionally painted with strokes of human creativity and intuition, is now experiencing an infusion of artificial intelligence. As AI's capabilities have expanded, so has its integration into the world of design, leading to a fusion of human artistry with machine precision.

AI's Brushstrokes in Design

- **Automated Layout Suggestions:** Tools like Canva harness AI algorithms to offer layout suggestions tailored to the user's content, ensuring visual harmony and aesthetic appeal without the need for manual adjustments.

- **Intelligent Color Palettes:** Deciding on a color scheme can be a nuanced task, balancing aesthetics with brand messaging. AI-driven tools can analyze images or content and suggest complementary color palettes, eliminating the guesswork and streamlining the design process.

- **Font Pairing:** Adobe's Sensei, among other tools, offers intelligent font pairing suggestions, matching typefaces in a manner that is visually cohesive and aligned with the design's intent.

- **Image Recognition and Enhancement:** AI can identify elements within images, allowing for targeted enhancements, automatic tagging, or even content-specific design suggestions.

- **Style Transfer:** Leveraging neural networks, designers can apply the artistic style of one image to another, leading to creative compositions that might have been time-consuming or impossible manually.

Implications and Potential

- **Efficiency Boost:** With AI handling some of the routine and repetitive tasks, designers can allocate more time to conceptualizing, strategizing, and executing complex design elements.

- **Bridging Skill Gaps:** For novices or non-designers, AI-powered tools can simplify the design process, allowing them to create professional-quality graphics without a steep learning curve.

- **Unleashing Creativity:** By offering suggestions and inspirations, AI can act as a digital muse, sparking creativity and encouraging designers to explore new design avenues.

- **Personalization at Scale:** For projects that require personalization, like marketing campaigns targeted at specific demographics, AI can automate the design variations while maintaining brand consistency.

Challenges and Considerations

- **Over-reliance on AI:** While AI can assist, over-reliance on its suggestions might stifle originality, leading to designs that lack a unique touch or become formulaic.

- **Ethical Concerns:** AI tools that leverage style transfer or auto-generate designs might raise questions about originality and copyright, especially if they closely emulate copyrighted works.

Envisioning the Future

The symbiosis of AI and graphic design is still in its nascent stages, but the trajectory is promising. As AI algorithms become more refined and designers grow more adept at harnessing their potential, the future of graphic design promises to be a tapestry woven with threads of human creativity and AI precision. The challenge and opportunity lie in ensuring that this fusion enhances the essence of design rather than overshadowing it.

These case studies illustrate the unexpected ways AI has impacted creative jobs. They show that AI is not just about automating manual labor or routine tasks; it's also about assisting with and even automating creative tasks. In the next section, we'll delve deeper into this idea, exploring the concept of the "middle layer" in AI applications and its importance for business opportunities.

The Middle Layer

As we've seen in the case studies, AI has made significant strides in automating tasks across various fields. However, the true potential of AI lies not just in automating tasks but in creating new opportunities for innovation and value creation. This is where the concept of the "middle layer" comes into play.

The "middle layer" refers to the space between the base AI technology and the end user application. It's the layer where raw AI capabilities are transformed into valuable and accessible tools for businesses and individuals. This layer is often overlooked, but it's crucial for unlocking the full potential of AI.

Consider the case of language models. At the base level, a language model is a complex technology that can generate human-like text. However, this raw capability is only directly useful for some businesses or individuals. They don't need a tool that generates random text; they need a tool that solves a specific problem or meets a particular need.

This is where the middle layer comes in. It's the layer where raw AI capabilities are packaged into specific applications, like a chatbot for customer service, a tool for drafting legal documents, or a system for generating news articles. These applications take the raw capabilities of the AI and apply them to specific tasks, creating value for businesses and individuals.

The middle layer is also where innovation and differentiation occur. While the base AI technology may be the same, the applications built on top of it can vary widely. Businesses can differentiate themselves by building unique applications that meet the needs of their customers, leveraging the capabilities of AI in new and innovative ways.

In the next section, we'll explore the future of AI evolution, discussing the potential for further advancements and the implications for businesses and society. As we'll see, the middle layer will play a crucial role in this future, bridging AI technology and its real-world applications.

The Future of AI Evolution

As we look to the future of AI evolution, we see a landscape filled with potential. The advancements we've seen are just the beginning, and there are many more opportunities for innovation and growth. In this section, we'll explore some potential future developments in AI and their implications for businesses and society.

One of the key areas of future development is the continued advancement of AI capabilities. As AI technology improves, we can expect to see even more sophisticated applications. For instance, language models may become better at understanding context and generating more nuanced and accurate text. This could open up new possibilities for AI in fields like journalism, law, and education—as mentioned in some of the previous case studies.

Another area of development is the expansion of AI into new fields. While AI has already made significant inroads into creative fields, there are many other areas where it has yet to make a significant impact. For instance, AI could be used in fields like health care to automate tasks like medical diagnosis or treatment planning. This could improve the efficiency and accuracy of healthcare services, leading to better patient outcomes.

The future of AI also holds potential challenges. As AI becomes more prevalent, data privacy, security, and ethical considerations will become increasingly important. Businesses and policymakers will need to navigate these challenges carefully to ensure the responsible use of AI.

Finally, the future of AI will likely involve a shift in the nature of work. As AI automates more tasks, the nature of human work will change. Jobs that involve routine tasks become less common, while jobs that involve complex problem-solving, creativity, and interpersonal skills become more important. This could lead to a shift in the skills in demand in the job market. A famous phrase I hear a lot when I speak at conferences is, "AI won't take your job, someone using AI will." A funny play on words, but nonetheless I believe it holds true at least in the transition period we are heading toward as a society. I don't believe people should be afraid of AI at all. Historically, we haven't seen mass layoffs due to advancements in technology at scale the way people have started talking about AI in recent years. I think of AI more as a tool. I laughed when people talked about "prompt-engineering" as a real job. I don't believe it will be a real job. I believe prompt engineering is a skill, much like "googling" is a skill in our world today. So instead of being afraid of AI, learn the skill. Play around with it. You will be amazed with the efficiencies you might find once you get used to having an artificial assistant. Here are two examples from my own life: AI has sped up my coding efficiency significantly, and when I speak at conferences, my entire slide deck is autogenerated from my synopsis of talking points. Life is getting faster and faster, we have to do our part to keep up.

In all of these areas of developments, which we have already started to see the beginning of, the middle layer will play a crucial role. It's in the middle layer that raw AI capabilities are transformed into valuable applications, and it's here that businesses can differentiate themselves and create value. By understanding and leveraging the middle layer, companies can position themselves for success in the future of AI.

When the development race of generative AI models (currently going on between OpenAI and the rest of the field) has played out, we will most likely have one or two main general models which can do most things with such high accuracy that spending millions training your own models won't be feasible for companies. I get a lot of questions from business owners about how they should train their company's custom language models, and my answer is always the same, "don't." In most general use cases in the world today, a fine-tuned model, perhaps with a RAG (Retrieval Augmented Generation, a fancy term for a custom vector database where the language model can extract custom insights from the company's own data), will perform immediately and won't cost much to develop. Most decent data scientists can spin up a prototype in a day or two. Meaning the company can spend the remaining time optimizing and improving the fine-tuned model's performance. The end result will, in the vast majority of cases, outperform a custom-trained model at a fraction of the cost.

Conclusion

Looking to the future, we expect to see further advancements in AI capabilities, the expansion of AI into new fields, and a shift in the nature of work. These developments hold great potential, but they also present challenges that must be navigated carefully.

In the end, the unexpected evolution of AI serves as a reminder of the transformative power of this technology. It challenges our assumptions, pushes the boundaries of what's possible, and opens up new possibilities for innovation and growth. As we continue to explore and harness the potential of AI, we can look forward to a future filled with exciting opportunities and challenges. So don't let the news headlines sway your world view. They are only trying to sell more newspapers. Stay positive, embrace the advancements and you will find your spot here in life much more enjoyable.

CHAPTER 5

The Business Opportunities of Today

As we explore the business opportunities presented by the current state of AI, it's essential to understand the context in which these opportunities exist. The rapid evolution of AI technology has opened up new avenues for businesses, transforming industries and redefining what's possible. Understanding and leveraging these opportunities is crucial for companies looking to stay competitive in today's digital age.

Artificial Intelligence, once a concept confined to science fiction, is now a reality reshaping the world as we know it. From self-driving cars to virtual assistants, AI is increasingly becoming a part of our everyday lives. But beyond these consumer-facing applications, AI also creates significant business opportunities across various industries.

The power of AI lies in its ability to process and analyze vast amounts of data at a speed and accuracy that far surpasses human capabilities. This ability can be leveraged in numerous ways, from automating routine tasks to generating insights from data, creating new products and services, and enhancing decision-making processes.

However, the potential of AI extends beyond simply improving efficiency and productivity. AI also opens up opportunities for innovation, enabling businesses to create new products and services, enter new markets, and redefine their business models. In this sense, AI is not just a tool for doing things better but also for doing better things.

© Jonas Bjerg 2024
J. Bjerg, *The Early-Career Professional's Guide to Generative AI*,
https://doi.org/10.1007/979-8-8688-0456-4_5

But to fully leverage these opportunities, businesses need to understand the landscape of AI, including the capabilities of current AI technologies, the trends shaping the AI field, and the challenges and ethical considerations associated with AI use. This understanding forms the foundation for identifying and capitalizing on AI's business opportunities.

In the following sections, we'll explore these opportunities, exploring how businesses can leverage AI in search, services, and startups. We'll also revisit the "middle layer" concept in AI applications, discussing its importance in creating business opportunities. As we navigate through these topics, we'll gain a deeper understanding of the business opportunities presented by the current state of AI.

The Shift in Search

One of the most significant shifts in the digital landscape over the past decade has been the evolution of search. Traditionally, search engines like Google have provided users with a list of links or options related to their search query. However, the nature of search is changing, which presents a significant business opportunity.

Today, most searches are not looking for multiple options but rather a single, accurate answer. Whether it's a question about the weather, a fact-check, or a complex query related to a specific field of knowledge, users increasingly expect direct and precise answers from their search queries. This shift in user expectations has been partly driven by the rise of digital assistants like Siri and Alexa, designed to provide direct answers to user queries.

This shift in search presents a significant opportunity for businesses. Providing direct, accurate answers to user queries can help companies to establish themselves as authoritative sources of information, build trust with users, and drive traffic to their websites. However, to capitalize on this opportunity, businesses need to understand and leverage the capabilities of AI.

AI, specifically natural language processing, plays a crucial role in this new era of search. AI algorithms can analyze and understand user queries, determine their intent, and generate accurate, contextually relevant responses. This capability is at the heart of digital assistants and is increasingly used by search engines to answer user queries directly.

However, the potential of AI in search extends beyond simply providing direct answers. AI can also be used to personalize search results, understand complex queries, and even predict user needs before they search. These capabilities can provide a more engaging and satisfying search experience for users, further enhancing the value of search for businesses.

In the next section, we'll delve deeper into the role of AI in various service industries, exploring how businesses can leverage AI to create new opportunities and enhance their services.

AI in Services

The service sector, encompassing a wide range of industries from health care to education, is another area where AI creates significant business opportunities. By automating tasks, generating insights, and enhancing service delivery, AI transforms how services are provided and consumed.

Health Care

The healthcare domain, with its intricate complexities and profound impact on human lives, is undergoing a transformational shift courtesy of artificial intelligence. While the essence of medical care remains deeply human, AI is enhancing the precision, efficiency, and reach of healthcare services, heralding a new era in medicine.

AI's Multifaceted Role in Health Care

- **Automated Patient Triage:** Hospitals and clinics receive an influx of patients daily. AI-driven systems can automatically assess the severity of a patient's symptoms and prioritize cases, ensuring that critical cases receive timely attention.

- **Medical Imaging Mastery:** Medical imaging, whether it's X-rays, MRIs, or CT scans, generates a wealth of visual data. AI algorithms, especially those using deep learning, can scrutinize these images pixel by pixel, detecting anomalies or signs of diseases like tumors, fractures, or infections. Tools like Google's DeepMind have showcased the potential of AI in diagnosing eye diseases through retinal scans with remarkable accuracy.

- **Data-Driven Diagnosis:** With access to vast patient datasets, AI can identify patterns, correlations, and outliers, assisting doctors in diagnosing diseases. By analyzing symptoms, medical histories, and even genetic data, AI can offer diagnostic suggestions or highlight potential areas of concern.

- **Predictive Care:** AI's predictive analytics capabilities can forecast patient trajectories. For instance, it can predict potential complications for hospitalized patients or the likelihood of readmission, allowing for pre-emptive interventions.

Implications and Potential

- **Enhanced Accuracy:** AI's ability to process vast amounts of data quickly can lead to more accurate diagnoses, reducing the chances of oversights or errors.

- **Efficiency and Cost Savings:** By automating routine tasks, AI can streamline healthcare operations, leading to quicker services and potential cost savings.

- **Personalized Treatment:** AI can analyze individual patient data to recommend treatments tailored to specific patient profiles, moving toward more personalized medicine.

- **Global Health Equity:** AI-powered diagnostic tools can be deployed in areas with limited access to medical experts, democratizing access to quality health care.

Challenges and Considerations

- **Data Privacy:** Handling medical data comes with stringent privacy requirements. Ensuring that AI systems are secure and compliant with regulations like HIPAA is paramount.

- **Ethical Concerns:** The use of AI in critical medical decisions raises ethical questions, especially in cases of misdiagnoses or errors.

- **Human Touch:** While AI can assist with diagnostics and predictions, human touch, empathy, and interpersonal communication remain irreplaceable in-patient care.

Looking Ahead

The amalgamation of AI and health care promises a future where medical care is not only more advanced but also more accessible and personalized. As research and innovations continue, the synergy between AI algorithms and medical expertise will deepen, driving the healthcare sector toward unprecedented horizons. The challenge lies in harnessing AI's potential responsibly, ensuring that technology enhances care without overshadowing the human essence of healing.

Legal Services

The world of law, with its intricate tapestry of statutes, precedents, and legalese, has traditionally been a domain marked by meticulous human scrutiny. Yet, the march of technology, spearheaded by artificial intelligence, is bringing transformative changes to the hallowed halls of legal practice. Let's delve into the myriad ways AI is redefining legal services.

AI's Multidimensional Role in Law

- **Revolutionizing Legal Research:** The labyrinth of legal texts, from case laws to statutes, demands rigorous research. AI-powered platforms can trawl through vast databases, pinpointing relevant legal references, precedents, or contradictions with unparalleled speed and precision. This not only speeds up the research process but also ensures a comprehensive review.

- **Contract Scrutiny:** Contracts, while standardized in structure, contain nuances that can have profound implications. AI systems can dissect contracts, highlighting clauses that deviate from the norm, potential legal risks, or even suggesting modifications to ensure compliance and protect clients' interests.

- **Document Review and Discovery:** In litigation, reviewing piles of documents to identify pertinent evidence is a colossal task. AI-driven tools can automate this process, flagging relevant documents, potential inconsistencies, or crucial information based on predefined parameters.

- **Predictive Analysis:** Some AI tools are venturing into legal predictive analytics, analyzing past case outcomes to gauge the likely trajectory of current cases. This can offer lawyers insights into potential verdicts or inform litigation strategies.

Implications and Advantages

- **Efficiency Boost:** By automating time-consuming tasks, lawyers can allocate more time to client consultations, strategy formulation, or court appearances, enhancing overall efficiency.

- **Enhanced Accuracy:** The sheer volume of legal data can lead to oversights. AI, with its meticulous data processing capabilities, reduces the margin for error.

- **Cost-Effective:** Automation can lead to cost savings for legal firms, potentially translating to more competitive billing rates for clients.

- **Data-Driven Insights:** AI's ability to identify patterns or correlations in legal data can offer lawyers unique insights, informing more strategic decision-making.

Challenges and Ethical Dimensions

- **Reliability and Accountability:** While AI tools can assist, relying solely on their recommendations without human oversight might lead to inaccuracies or misjudgments.

- **Data Privacy and Security:** Legal documents often contain sensitive information. Ensuring that AI systems handle this data with the utmost security and adhere to confidentiality norms is crucial.

- **Job Displacement Concerns:** As AI takes over routine tasks, there are concerns about its impact on junior legal roles traditionally responsible for research or document review.

The Road Ahead

The melding of AI with legal services is reshaping the legal landscape. While AI tools offer unprecedented advantages, a balanced approach, where AI complements human expertise, seems to be the most promising path forward. As AI continues to evolve, its integration with legal services will deepen, forging a future where law is not just about human judgment but also about harnessing the power of data-driven insights.

Education

Education, a cornerstone of personal and societal development, is currently in the midst of a digital renaissance. Central to this transformation is artificial intelligence, which introduces novel methodologies and tools to the traditional classroom setting. From personalized learning trajectories to smart content creation, AI is redefining the contours of education and the student experience.

AI's Multifaceted Contributions to Education

- **Personalized Learning Pathways:** Every student is unique, with individual learning styles, strengths, and areas of challenge. AI systems can analyze student performance data, learning preferences, and engagement patterns to curate personalized study plans, ensuring that each student receives instruction tailored to their needs.

- **Automated Grading:** Grading, especially for large classes or standardized tests, can be a time-consuming endeavor. AI-driven platforms can automatically grade multiple-choice, fill-in-the-blank, and even certain open-ended questions, freeing educators to focus on more complex evaluative tasks and interactive teaching.

- **AI-Powered Tutors:** For subjects like mathematics, science, or languages, AI-powered tutoring systems can assist students outside of classroom hours. These systems can provide instant feedback, adapt problems to the student's proficiency level, and even simulate human-like interactions, making learning more engaging.

- **Smart Content Creation:** AI tools can generate customized reading material, quizzes, or interactive content based on the curriculum and the learner's proficiency level. This dynamic content adjusts in real time as the student progresses.

- **Learning Analytics:** AI-driven analytics can provide insights into student engagement, areas of struggle, or potential risks of dropping out, enabling educators to intervene proactively.

Implications and Potential

- **Enhanced Student Engagement:** With content tailored to their needs and instant feedback mechanisms, students are likely to find learning more engaging and fulfilling.

- **Efficiency and Scalability:** Automated grading and content generation can cater to larger student groups without compromising on individual attention, making quality education more scalable.

- **Data-Driven Insights:** With AI's capability to analyze vast amounts of student data, educators can gain deeper insights into classroom dynamics, learning outcomes, and areas for pedagogical improvement.

Challenges and Ethical Considerations

- **Data Privacy:** Handling student data comes with the responsibility of ensuring privacy and security. Schools and ed-tech platforms must ensure compliance with regulations and protect student information.

- **Equity Concerns:** There's a risk of AI systems perpetuating biases if not carefully trained, potentially putting certain groups of students at a disadvantage.

- **Human Interaction:** While AI can simulate interactions, the irreplaceable value of human mentorship, peer interactions, and collaborative learning must be recognized.

The Road Ahead

The synergy between AI and education is opening up vistas of possibilities. As we advance, we can expect even more immersive learning experiences, with virtual reality, augmented reality, and AI merging to create rich, multidimensional educational platforms. However, as with any technological integration, the key lies in balance—using AI to enhance, not replace, the human essence of teaching and learning.

These examples illustrate the potential of AI in the service sector. By automating tasks, generating insights, and enhancing service delivery, AI can help businesses provide better services, improve efficiency, and create new opportunities. However, to fully leverage these opportunities, companies need to understand the capabilities of AI, the needs of their customers, and the regulatory and ethical considerations associated with AI use.

In the next section, we'll revisit the "middle layer" concept in AI applications, discussing its importance in creating business opportunities and how businesses can leverage this layer to create value.

The Importance of the Middle Layer

As we delve deeper into the business opportunities presented by AI, it's essential to revisit the concept of the "middle layer" in AI applications. This layer, which sits between the base AI technology and the end user application, plays a crucial role in transforming raw AI capabilities into practical, value-creating applications.

The middle layer is where the raw capabilities of AI are packaged into specific applications that solve real-world problems or meet specific needs. For instance, a language model's ability to generate human-like text becomes a practical application when it's used to create a customer service chatbot, a content generation tool, or a system for automating administrative tasks.

This transformation process is not just about technical implementation. It also involves understanding the needs of the end users, the context in which the application will be used, and the business model that will support the application. This requires a blend of technical expertise, business acumen, and user-centric design—skills often found in product managers, business strategists, and user experience designers.

The middle layer is also where differentiation and innovation occur. While the base AI technology may be the same, the applications built on top of it can vary widely. By creating unique applications that meet the needs of their customers, businesses can differentiate themselves from their competitors and create a defendable position in the market.

Moreover, the middle layer is where businesses can create value and generate revenue. Whether it's through selling AI-powered products or services, using AI to enhance existing offerings, or leveraging AI to create operational efficiencies, the middle layer is where the economic value of AI is realized.

In the next section, we'll explore the future of AI startups and discuss why there may be other keys to success than developing proprietary AI models. Instead, we'll argue that the real opportunity lies in building upon existing models and focusing on application and implementation.

The Future of AI Startups

As we look to the future of AI startups, a key question emerges: What is the path to success in a field that is rapidly evolving and increasingly competitive? The answer may not lie in developing proprietary AI models. Instead, the real opportunity may lie in building upon existing models and focusing on application and implementation.

The development of AI models is a complex, resource-intensive process. It requires significant expertise, vast amounts of data, and substantial computational resources. Moreover, AI research is highly competitive, with tech giants like Google, Facebook, and OpenAI often leading the way in developing new models.

Given these challenges, it may not be feasible or strategic for most startups to develop their own AI models. Instead, a more viable approach may be to build upon existing models that are openly available. For instance, models like GPT-4 and BERT, developed by OpenAI and Google, respectively, are available for businesses to use and build upon.

Building upon existing models allows startups to leverage the cutting-edge capabilities of these models without the need to develop them from scratch. This can significantly reduce the time, cost, and risk associated with developing proprietary models, allowing startups to focus their resources on application and implementation.

Focusing on application and implementation means using AI to solve real-world problems or meet specific needs. This involves understanding the needs of the end users, designing applications that meet these needs, and implementing these applications in a way that creates value for the users and the business. This is where the middle layer comes into play, transforming raw AI capabilities into practical, value-creating applications.

Moreover, focusing on application and implementation allows startups to create a defendable position in the market. While AI models may be openly available, the applications built on top of them can be unique and differentiated. By creating unique applications that meet the needs of their customers, startups can differentiate themselves from their competitors and create a defendable position in the market.

In the next section, we'll delve into the role of AI tools like Dall-E or Co-pilot, discussing how these tools can be used to create business opportunities and why they're essential for the future of AI.

The Role of AI Tools

As we explore the business opportunities AI presents, it's important to consider the role of AI tools. These tools, which include AI-powered systems like Dall-E and Co-pilot, are transforming the way we work and creating new opportunities for businesses.

Dall-E, developed by OpenAI, is an AI system that generates images from textual descriptions. It can create unique, never-before-seen images based on a wide range of prompts, from everyday objects to fantastical creatures. This capability opens up new possibilities in fields like graphic design, advertising, and entertainment, where businesses can leverage Dall-E to create original visual content.

Co-pilot, also developed by OpenAI, is an AI system that assists with coding. It provides suggestions for completing lines or blocks of code, helping developers write more efficiently and effectively. This capability can be leveraged in fields like software development and data analysis, where businesses can use Co-pilot to enhance productivity and reduce the time and effort required to write code.

These AI tools illustrate the transformative power of AI. By automating tasks, enhancing productivity, and enabling new capabilities, these tools are creating new opportunities for businesses. They're also reshaping the nature of work, shifting the focus from routine tasks to more complex, creative, and strategic tasks.

However, to fully leverage these opportunities, businesses need to understand how to use these tools effectively. This involves understanding the capabilities and limitations of the tools, integrating them into existing workflows, and training employees to use them effectively. It also involves navigating the ethical and regulatory considerations associated with AI use.

The future of AI is filled with potential, but realizing this potential requires a deep understanding of AI and a strategic approach to leveraging its capabilities.

Conclusion

The business opportunities presented by AI are vast and varied. However, realizing these opportunities requires a deep understanding of AI, a strategic approach to leveraging its capabilities, and a commitment to navigating the challenges and ethical considerations associated with AI use. As we continue to explore and harness the potential of AI, we can look forward to a future filled with exciting opportunities and challenges.

CHAPTER 6

The Future of AGI

The rapid advancements in AI technology, coupled with the increasing integration of AI into our everyday lives, are setting the stage for a future that is both exciting and challenging.

The future of AI holds immense potential. From the continued evolution of language models to the development of Artificial General Intelligence (AGI), the capabilities of AI are set to expand in ways that could transform various fields, from health care and education to transportation and entertainment.

However, this future is challenging. As AI becomes more prevalent and powerful, issues such as data privacy, security, and ethical considerations will become increasingly important. The impact of AI on jobs and society will also need to be carefully managed to ensure a future that is inclusive and equitable.

Moreover, the future of AI will likely involve a shift in the nature of work and the skills that are in demand. As AI automates more tasks, jobs that involve routine tasks may become less common, while jobs that involve complex problem-solving, creativity, and interpersonal skills may become more important.

In this chapter, we'll delve deeper into these topics, exploring the potential advancements, challenges, and implications of the future of AI. We'll discuss the potential of language models, the concepts of self-improvement and algorithmic improvement, the development of AGI, and the impact of AI on jobs and society. We'll also explore the potential devaluation of AI-generated artworks and the concept of the AI gold rush.

© Jonas Bjerg 2024
J. Bjerg, *The Early-Career Professional's Guide to Generative AI*,
https://doi.org/10.1007/979-8-8688-0456-4_6

As we navigate through these topics, we'll gain a deeper understanding of the future of AI, providing insights that can help us prepare for and shape this future. Let's begin this journey by exploring the potential of language models in the next section.

The Potential of Language Models

Language models, such as GPT-4 developed by OpenAI, have been at the forefront of AI advancements in recent years. These models, trained on vast amounts of text data, can generate human-like text that can be remarkably coherent, creative, and contextually relevant. As we look to the future, the potential of these language models is vast and exciting.

One area where we can expect to see significant advancements is in the understanding and generation of contextually nuanced text. Current language models are already quite adept at this, but there's still room for improvement. Future models could become even better at understanding the nuances of language, including things like sarcasm, humor, and cultural references. This could make interactions with AI more natural and engaging, opening up new possibilities for AI in fields like customer service, entertainment, and education.

Another potential advancement area is the ability of language models to interact with other types of data. Currently, language models primarily deal with text data. Still, future models could be trained to understand and generate different data types, such as images, audio, or even structured data like spreadsheets. This could enable more integrated and versatile AI applications, such as a virtual assistant that can understand and generate both text and images or an AI analyst that can understand and generate insights from structured data.

The potential of language models also extends to their ability to learn and improve over time. Future models could become better at learning from their interactions, adapting their responses based on feedback, and improving their performance over time. This could make AI more responsive and effective, enhancing its value for businesses and individuals.

However, the potential of language models also brings challenges. As these models become more powerful and pervasive, issues like data privacy, misinformation, and the digital divide will become increasingly important. Navigating these challenges will be crucial for realizing the potential of language models and ensuring a future that is inclusive, equitable, and beneficial for all.

In the next section, we'll delve deeper into the concepts of self-improvement and algorithmic improvement in AI, discussing how these could shape the future of AI.

Self-improvement and Algorithmic Improvement

As we look to the future of AI, two concepts stand out as potential game-changers: self-improvement and algorithmic improvement. These concepts, while still largely theoretical, could significantly enhance the capabilities of AI and shape its future trajectory.

Self-improvement

Self-improvement in AI refers to the ability of an AI system to learn from its experiences and improve its performance over time. This is already a feature of many AI systems today, which use techniques like reinforcement learning to adapt and optimize their behavior based on feedback.

However, the potential for self-improvement in AI goes beyond what we see today. Future AI systems could become much more adept at learning from their experiences, adapting their behavior in more sophisticated ways, and even devising their own strategies for learning and improvement. This could lead to AI systems that are more autonomous, adaptable, and effective.

Algorithmic Improvement

Algorithmic improvement refers to the ability of an AI system to improve its own algorithms, essentially becoming better at becoming better. This concept, or recursive self-improvement, is a key idea in discussions about Artificial General Intelligence (AGI).

In theory, an AI system capable of algorithmic improvement could undergo a rapid process of self-enhancement, leading to exponential growth in its capabilities. This could result in an AI system that quickly surpasses human intelligence, a scenario often referred to as the "intelligence explosion" or "singularity."

However, the concept of algorithmic improvement also raises significant challenges and risks. A self-improving AI could become uncontrollable, leading to outcomes that are harmful to humanity. This is known as the alignment problem, which we'll discuss in more detail in the next section.

In conclusion, the concepts of self-improvement and algorithmic improvement could significantly shape the future of AI. While they hold great potential, they also present significant challenges that must be carefully managed. As we continue to explore the future of AI, these concepts will be crucial areas of focus.

Artificial General Intelligence (AGI) and the Alignment Problem

Artificial General Intelligence (AGI), often referred to as "strong AI," is a type of artificial intelligence that has the ability to understand, learn, and apply knowledge across a wide range of tasks at a level equal to or beyond that of a human. Unlike narrow AI, designed to perform specific tasks, AGI can perform any intellectual task a human can.

The development of AGI is a long-standing goal in AI, and it represents a significant leap forward in AI capabilities. However, the path to AGI is fraught with challenges, both technical and ethical. One of the most important of these is the alignment problem.

The alignment problem refers to the challenge of ensuring that AGI's goals and behaviors align with human values and interests. As AGI would be highly autonomous and capable of self-improvement, it's crucial that its actions result in outcomes that are beneficial to humanity.

However, specifying human values in a way that an AGI can understand and follow is a complex task. Human values are often vague, context-dependent, and subject to change, making them difficult to encode in an AI system. Moreover, there's the risk that an AGI could interpret its goals in unintended ways, leading to harmful outcomes, a scenario often referred to as "perverse instantiation."

Addressing the alignment problem will require significant research and careful design. It will also require ongoing oversight and control mechanisms to ensure that AGI remains aligned with human values as it learns and evolves. This is a significant challenge, but it's crucial for ensuring that the development of AGI results in outcomes that benefit humanity.

In the next section, we'll explore the potential impact of AI on jobs and society, discussing issues like mass layoffs, the shift in job skills, and the likely need for measures like Universal Basic Income (UBI).

The Impact on Jobs and Society

As AI advances and becomes more integrated into our everyday lives, it will inevitably significantly impact jobs and society. While these changes can bring about many benefits, they also present challenges that need to be carefully managed.

Impact on Jobs

The proliferation of artificial intelligence in various sectors is akin to a technological tidal wave, reshaping industries and fundamentally altering the nature of work. While the narrative around AI often veers toward job displacement, it's essential to view this transformation through a broader lens, recognizing both the challenges and opportunities AI presents in the job market.

AI and Job Displacement

- **Routine Task Automation:** Jobs that revolve around repetitive, structured tasks—be it data entry, basic customer service, or certain manual labor tasks—are particularly susceptible to automation. AI systems, with their ability to learn and execute tasks with consistent precision, can perform these jobs often faster and more accurately than humans.

- **Specialized Job Automation:** Beyond routine tasks, AI is also venturing into specialized domains. For instance, diagnostic AI tools in health care or AI-driven financial analysis tools might reduce the demand for entry-level positions in these sectors.

The Silver Lining: Job Creation and Evolution

- **Emergence of New Roles:** While AI might render certain jobs obsolete, it's also paving the way for novel roles. Jobs related to AI development, maintenance, and oversight, such as AI trainers, algorithm specialists, or AI ethicists, are on the rise.

- **Augmentation of Existing Jobs:** Instead of replacing humans, AI can act as a powerful tool in the hands of professionals. For instance, doctors can use AI for preliminary diagnostics, allowing them to focus on treatment plans or complex cases. Similarly, designers might use AI for basic layouts, reserving their time for creative conceptualization.

- **Skill Upgradation:** As routine tasks get automated, employees can be upskilled to take on more complex roles within the same organization, leading to career progression and diversification.

Broader Implications

- **Economic Shifts:** As industries embrace AI, there might be economic shifts, with some sectors booming while others contract. This can have ramifications on job availability in various regions or sectors.

- **Educational Evolution:** The rise of AI necessitates a rethinking of educational curricula. Emphasizing skills that AI struggles to replicate—creativity, emotional intelligence, critical thinking—becomes paramount.

- **Social and Ethical Considerations:** The potential displacement of jobs by AI raises pressing social questions about income disparities, job security, and societal structures.

In Conclusion: The dance between AI and the job market is intricate, with both challenges and opportunities in its wake. While the fear of job loss is tangible and valid, it's equally vital to recognize AI's potential to usher in a new era of work marked by collaboration between humans and machines. The future might not be about AI vs. humans, but rather AI and humans working in tandem to achieve goals neither could accomplish alone.

Shift in Job Skills

The metamorphosis of the job market under the influence of artificial intelligence is not just about the roles themselves but also about the skills that underpin them. As AI seamlessly integrates into various sectors, it brings about a recalibration of the skills that are deemed essential. Let's delve into how the rise of AI is reshaping the skills landscape and what it means for the future workforce.

The Onset of Technical Proficiencies

- **AI and Machine Learning Mastery:** As businesses and industries harness the power of AI, there's a burgeoning demand for professional's adept in machine learning algorithms, neural networks, and AI system development.

- **Data Science and Analysis:** Data is the lifeblood of AI. Professionals skilled in data collection, processing, and analysis are pivotal in training robust AI systems. This encompasses not just data crunching but also understanding the implications and nuances of the data.

- **Robotics and Automation:** With the proliferation of AI in manufacturing and logistics, skills related to robotics, robot maintenance, and automated systems are gaining prominence.

The Resurgence of Inherently Human Skills

- **Creativity:** While AI can generate art, music, or even write articles, the depth, nuance, and originality of human creativity remain unparalleled. Jobs that revolve around ideation, design, and artistic creation are likely to thrive.

- **Critical Thinking:** The ability to assess situations, draw from diverse knowledge sources, and make informed decisions—especially in unfamiliar scenarios—is a forte of human cognition. In an era of information overload, critical thinkers who can discern quality information from noise are indispensable.

- **Emotional Intelligence:** Human interactions, empathy, and understanding of emotional nuances are arenas where AI lags. Professionals skilled in interpersonal relations, counseling, HR, and leadership roles will find their skills in high demand.

- **Problem-Solving:** While AI can solve predefined problems, tackling novel challenges or developing innovative solutions requires human ingenuity.

Adaptable and Lifelong Learning

- **Continuous Learning:** Given the rapid pace of technological advancements, the ability to learn, unlearn, and relearn becomes crucial. Adaptable individuals who proactively upgrade their skills will be better positioned in the evolving job market.

- **Interdisciplinary Knowledge:** As boundaries between fields blur, individuals who can bridge disciplines—combining, say, biology with AI or design with technology—will be highly sought after.

In Conclusion: The AI-driven shift in job skills paints a future where technical proficiency and human-centric skills coalesce. It underscores a future where the most valued professionals are not just those adept at working with machines but those who bring a blend of technical know-how and human touch. As we navigate this transition, fostering a culture of continuous learning and valuing diverse skill sets will be paramount.

Universal Basic Income (UBI)

The march of artificial intelligence, while laden with promises of efficiency and innovation, also brings forth profound socioeconomic implications. Central to the debate on the future of work in the AI era is the concept of Universal Basic Income (UBI), a radical economic proposition that is increasingly gaining traction. Let's delve deeper into UBI, its motivations, potential benefits, challenges, and the interplay with AI-driven transformations.

Motivations for UBI in an AI-Dominated World

- **Mitigating Job Displacement:** As AI automates a spectrum of tasks, from routine to specialized, there's a palpable fear of widespread job displacement. UBI acts as a financial buffer, ensuring individuals can sustain themselves even in periods of unemployment.

- **Income Inequality:** AI's rise might lead to increased profits for tech giants and those at the forefront of AI innovations, potentially widening income disparities. UBI aims to redistribute wealth and reduce this gap.

- **Economic Stability:** By ensuring a steady income for all, UBI can bolster consumer spending, acting as a stabilizing force for economies, especially in times of downturns.

Potential Benefits of UBI

- **Reduced Poverty:** With a guaranteed income, UBI can drastically reduce poverty levels, ensuring every citizen has the means to cover basic necessities.

- **Encouragement of Entrepreneurship:** Freed from financial precarity, individuals might be more inclined to pursue entrepreneurial endeavors, take creative risks, or invest in personal and professional development.

- **Simplified Welfare Systems:** UBI could replace or streamline complex welfare programs, reducing bureaucratic overhead and ensuring more direct benefit to citizens.

- **Enhanced Job Flexibility:** With a financial safety net, individuals might explore part-time work, freelancing, or roles that align more with their passions rather than financial compulsions.

Challenges and Criticisms of UBI

- **Economic Viability:** The foremost critique of UBI revolves around its funding. The significant financial outlay required raises questions about tax structures, national budgets, and the long-term sustainability of such a program.

- **Potential Inflation:** There are concerns that UBI could lead to inflation, with increased demand driving up prices, thereby eroding the very purchasing power UBI aims to provide.

- **Work Ethic Concerns:** Critics argue that a guaranteed income might disincentivize work, leading to reduced productivity and a potential societal shift in attitudes toward employment.

UBI in the Context of AI

- **AI-Driven Productivity Gains:** As AI boosts productivity, some argue that the resultant economic gains could fund UBI, transforming potential job losses into a societal advantage.

- **Transition Aid:** As the workforce navigates the transition from traditional roles to jobs of the future, UBI could provide financial stability, allowing individuals the time and resources to retrain or adapt.

The discourse on Universal Basic Income, especially in the context of AI, is multifaceted and evolving. It embodies society's efforts to grapple with the rapid technological transformations and their ripple effects. Whether

UBI is the panacea for the challenges of the AI epoch or just a part of a broader solution remains to be seen. What's certain is the need for proactive, inclusive, and forward-thinking policies as we step into a future shaped by AI.

The Devaluation of AI-Generated Artworks

As AI continues to evolve and expand its capabilities, it's beginning to encroach on areas once dominated by human intellect and creativity, prominently including the realm of artistic expression—areas once thought to be uniquely human.

With AI's foray into generating paintings, music, poetry, and beyond, the intrinsic value and definition of art are under scrutiny. This technological advancement has led to AI-created artworks that rival those of humans in appearance and have fetched substantial sums at auctions, prompting a reevaluation of what constitutes valuable art.

Art, historically revered for its aesthetic appeal, is also appreciated for the human ingenuity, emotional depth, and intentionality that fuel its creation. These elements contribute to the art's unique story and the connection it fosters between the creator and the audience. AI, in contrast, operates devoid of personal experiences, emotions, or preferences, merely executing programmed algorithms to produce art. This fundamental difference raises questions about the significance and worth of AI-generated art, which, though technically impressive, may lack the profound human experiences and narratives that often define and enrich art.

The potential devaluation of AI-generated artworks might stem from their lack of human connection, which could diminish their attractiveness and worth to art enthusiasts and collectors who seek depth and authenticity. Despite this, the integration of AI in art opens up unprecedented avenues for creativity and innovation. By serving as a tool for human artists, AI can enhance artistic exploration and creation, leading to the emergence of new forms of art.

An example of new forms of art could be how people have used AI to merge together instruments creating entirely new sounds. I went to school with a guy at Stanford who had trained a model to create new Kanye West songs. He simply gave the model a topic, and it created the melody, lyrics, and everything. I'm no expert, but it sounded like it was made by Kanye West to me. I don't mention this to say that people should use this to copy someone's style, but instead to show how AI technology enables people to be creative in new ways.

Furthermore, AI democratizes the artistic process, making art creation more accessible to those without traditional artistic skills, thus broadening the scope of who can participate in art-making.

In conclusion, the impact of AI on art is complex and uncertain. While it could lead to devaluing AI-generated art, it could also open up new possibilities for creativity and expression. As we continue to explore the future of AI, the intersection of AI and art will be an essential area to watch.

Conclusion

The potential of AI is vast and exciting. From the continued evolution of language models to the development of Artificial General Intelligence (AGI), the capabilities of AI are set to expand in ways that could transform various fields and aspects of our lives.

In the end, the future of AI is filled with potential, but realizing this potential requires a thoughtful and proactive approach. As we continue to explore and harness the potential of AI, we can look forward to a future filled with exciting opportunities and challenges. The journey toward that future starts with understanding and engaging with AI today.

CHAPTER 7

Navigating the AI Landscape

The rapid advancements in AI technology, coupled with the increasing integration of AI into our everyday lives, are creating an exciting and challenging landscape.

Navigating the AI landscape can be challenging. Individuals and businesses each have their own unique hurdles and opportunities. The landscape goes beyond just understanding the capabilities and limitations, identifying opportunities for leveraging AI, and managing the risks and challenges associated with AI use.

It involves technical knowledge, strategic thinking, ethical considerations, and a commitment to continuous learning and adaptation.

For businesses, navigating the AI landscape involves

- Integrating AI into their operations
- Developing new AI-based products or services
- Even transforming their business model to leverage the opportunities presented by AI

It can also involve managing the impact of AI on their workforce, customers, and competitive environment.

© Jonas Bjerg 2024
J. Bjerg, *The Early-Career Professional's Guide to Generative AI*,
https://doi.org/10.1007/979-8-8688-0456-4_7

For individuals, navigating the AI landscape can involve

- Understanding how AI impacts their work and daily life

- Developing skills in demand in an AI-driven world

- Learning how to interact effectively with AI systems

Additionally, it can involve understanding AI's ethical and societal implications and advocating for responsible and equitable AI use.

This chapter delves deeper into these topics, providing insights and guidance on navigating the AI landscape. We'll discuss the importance of understanding AI models, strategies for building a defensible position, the benefits of leveraging openly available models, and how to prepare for disruption due to AI. We'll also provide practical tips and tricks for prompt engineering with ChatGPT, a powerful AI model developed by OpenAI.

As we navigate through these topics, we'll gain a deeper understanding of the AI landscape, providing insights that can help us leverage the potential of AI and navigate its challenges. Let's begin this journey by exploring the world of AI models in the next section.

Understanding AI Models

AI models form the backbone of AI applications. They are the mathematical structures that learn from data and make predictions or decisions. Understanding these models, their capabilities, and their limitations is crucial to navigating the AI landscape.

AI models come in many forms, from simple linear regression models to complex deep learning models. Each model type has strengths and weaknesses and is suited for different tasks. For example, deep learning models are excellent at handling complex tasks like image recognition and natural language processing, but they require large amounts of data and computational resources.

One of the most significant advancements in AI in recent years has been the development of large language models like GPT-4. These models are trained on vast amounts of text data and can generate human-like text that is remarkably coherent and contextually relevant. They have been used in various applications, from customer service bots to creative writing tools.

However, while these models are powerful, they also have limitations. They need help understanding the text they generate; sometimes, they produce incorrect or nonsensical responses, and they can be sensitive to the input prompts given. Understanding these limitations is crucial for effectively using these models and managing their risks.

In addition to understanding individual AI models, it's essential to understand how models can be combined and integrated into larger systems. For example, a virtual assistant might use a language model to understand and generate text, a speech recognition model to convert speech to text, and a speech synthesis model to convert text to speech. Understanding how these models work together can help design and troubleshoot AI systems.

Building a Defensible Position

Building a defendable position is crucial as the AI landscape becomes increasingly competitive. This involves developing unique capabilities or assets that give you a competitive advantage and are difficult for others to replicate.

One way to build a defensible position is through technological superiority. This could involve developing advanced AI models, proprietary algorithms, or unique data assets. However, achieving technological superiority in AI is challenging due to the rapid pace of AI research and the widespread availability of AI technologies.

Another way to build a defensible position is through strategic positioning. This could involve targeting a specific niche, building strong customer relationships, or developing a unique brand. For example, a company might build a defendable position by becoming the go-to provider of AI solutions for a specific industry.

A third way to build a defensible position is through organizational capabilities. This could involve developing a robust AI team, cultivating an AI-friendly culture, or establishing effective AI governance practices. Organizational capabilities can be a powerful source of competitive advantage, as they are built up over time and are difficult for others to replicate.

Building a defendable position also involves managing risks. This includes technical risks, such as the risk of AI models failing or being misused, and strategic risks, such as competitors catching up. Effective risk management can protect your position and ensure the long-term success of your AI initiatives.

Leveraging Openly Available Models

In the rapidly evolving field of AI, leveraging openly available models can provide significant advantages. These models, which are often developed by leading AI research groups and shared with the broader community, can provide a powerful starting point for developing AI applications.

One of the most prominent platforms for openly available AI models is Hugging Face. Hugging Face hosts a wide range of models for tasks like text classification, named entity recognition, and text generation. These models are pre-trained on large datasets, meaning they have already learned useful features and patterns that can be applied to new tasks.

Leveraging these openly available models can save significant time and resources. Instead of developing and training a model from scratch, you can start with a pre-trained model and fine-tune it for your specific task.

This process, known as transfer learning, can often achieve high performance with less data and computational resources.

In addition to saving time and resources, leveraging openly available models can also provide access to state-of-the-art AI technologies. Many models on platforms like Hugging Face are based on the latest research and incorporate advanced techniques like transformer architectures and attention mechanisms.

However, while leveraging openly available models can provide significant advantages, it's also important to understand their limitations. These models are not a silver bullet and may only be suitable for some tasks. They also require careful handling to avoid issues like overfitting, bias, and misuse.

Preparing for Disruption

The rise of AI is set to disrupt various fields, from health care and education to finance and entertainment. While this disruption can bring many benefits, it also presents challenges that must be carefully managed.

For businesses, preparing for AI disruption involves

- Understanding how AI can impact their industry

- Identifying opportunities for leveraging AI

- Developing strategies to manage the risks associated with AI

This might involve integrating AI into their operations, developing new AI-based products or services, or even transforming their business model to leverage the opportunities presented by AI.

For individuals, preparing for AI disruption involves understanding how AI can impact their job and developing the skills that will be in demand in an AI-driven world. This might involve learning about AI and related fields like data science and machine learning, developing skills in areas that are difficult for AI to replicate, such as creativity and emotional intelligence, and learning how to interact effectively with AI systems.

Preparing for AI disruption also involves understanding and advocating for AI's ethical and societal implications. This includes issues like data privacy, algorithmic fairness, and the impact of AI on jobs and income distribution. By understanding these issues and advocating for responsible and equitable AI use, individuals and businesses can help shape the future of AI in a way that is beneficial for all.

Boosting Productivity with AI

AI has the potential to boost productivity in various fields significantly. By automating routine tasks, improving decision-making, and enabling new capabilities, AI can help individuals and businesses achieve more with less effort.

One way that AI can boost productivity is through automation. AI can automate a wide range of tasks, from data entry and scheduling to customer service and content creation. By automating these tasks, AI can free up time for individuals and businesses to focus on more complex, creative, and strategic tasks.

AI can also boost productivity by improving decision-making. AI can analyze large amounts of data, identify patterns and trends, and make predictions, helping individuals and businesses make more informed decisions. For example, AI can help businesses optimize their operations, target their marketing efforts, and improve their products and services.

In addition to automating tasks and improving decision-making, AI can also enable new capabilities that boost productivity. For example, AI can enable real-time language translation, personalized recommendations, and advanced analytics, among other things.

However, while AI can provide significant productivity benefits, managing the risks and challenges associated with AI use is also important. This includes technical challenges, such as the complexity of developing and maintaining AI systems, as well as ethical and societal challenges, such as the impact of AI on jobs, privacy, and fairness.

Prompt Engineering with ChatGPT

ChatGPT is a powerful language model developed by OpenAI that can generate human-like text. It can be used for a wide range of tasks, from drafting emails and writing articles to answering questions and creating conversational agents. However, to get the most out of ChatGPT, it's important to understand how to prompt it effectively.

Prompting ChatGPT involves providing it with an input text that it uses to generate an output text. The input text, or prompt, can significantly influence the output text, so crafting effective prompts is a crucial skill.

One approach to prompt engineering with ChatGPT is the CIDO method, which stands for Context, Instruction, Details, and Output.

- **Context** refers to the background information that helps ChatGPT understand the situation.

- **Instruction** is a clear command that tells ChatGPT what to do.

- **Details** are additional information that can help guide ChatGPT's response.

- **Output** is the desired result that you want from ChatGPT.

For example, if you want ChatGPT to help draft an email to a client, you might use the following prompt:

Context: "You are my assistant, and we have been working with a client named Alex on a project."

Instruction: "Draft an email"

Details: "The email should update Alex on our progress, ask for a meeting next week, and express appreciation for their patience."

Output: "Dear Alex, ..."

By using the CIDO method, you can craft effective prompts that guide ChatGPT to produce the desired output.

In addition to the CIDO method, there are other strategies for prompt engineering with ChatGPT. These include experimenting with different prompt styles, using explicit instructions, and iterating on prompts based on the outputs.

Conclusion

Several key insights emerge as we conclude this exploration of navigating the AI landscape. First, understanding AI models, their capabilities, and their limitations is crucial to leveraging AI effectively. This understanding can help guide the development of AI applications and the management of AI risks.

Second, building a defendable position in the AI landscape is crucial for achieving long-term success. This involves developing unique capabilities or assets, managing risks, and staying adaptable in the face of rapid technological change.

Third, leveraging openly available AI models can provide significant advantages, saving time and resources, providing access to state-of-the-art AI technologies, and accelerating AI development.

Fourth, preparing for AI disruption involves understanding how AI can impact various fields, developing the skills that will be in demand in an AI-driven world, and advocating for the ethical and societal implications of AI.

Fifth, AI has the potential to significantly boost productivity by automating tasks, improving decision-making, and enabling new capabilities. However, realizing these productivity benefits requires careful management of the risks and challenges associated with AI use.

Finally, prompt engineering with ChatGPT is a valuable skill for interacting effectively with this powerful AI model. By using methods like CIDO and experimenting with different prompt styles, you can guide ChatGPT to produce the desired outputs.

In conclusion, navigating the AI landscape is a complex but rewarding journey. It requires a deep understanding of AI, a strategic approach to leveraging its capabilities, and a commitment to continuous learning and adaptation. As we continue to explore and harness the potential of AI, we can look forward to a future filled with exciting opportunities and challenges. The journey toward that future starts with understanding and engaging with AI today.

CHAPTER 8

The Early Career Professional's Future With AI

The rapid advancements in AI technology, coupled with the increasing integration of AI into our everyday lives, are creating countless exciting opportunities and challenges.

The future of AI is not just about technology. It's also about people. It's about how AI will impact our jobs, our education, our daily lives, our rights, and our role in society. It's about how we can prepare for an AI-driven future and how we can shape this future in a way that aligns with our societal values and aspirations.

AI may seem like a distant and abstract concept for many. Especially before reading a book like this. The black box approach, often used to explain AI in general, instead of explaining the concepts behind the technology is a growing problem. Many publications and media curators seem to believe this approach is acceptable, presumably "until" AI is widely implemented and used. But the reality is that AI is already here, and it's becoming more pervasive with each passing day. From the recommendations we receive on streaming platforms, to the virtual assistants on our smartphones, to the predictive text features on our keyboards, AI is increasingly becoming a part of our daily lives. And simply

© Jonas Bjerg 2024
J. Bjerg, *The Early-Career Professional's Guide to Generative AI*,
https://doi.org/10.1007/979-8-8688-0456-4_8

throwing every type of AI into the same black box and saying that's enough of an explanation needed in each case, is only increasing the scare factor of the technology. The fact of the matter is that there are countless types of AI each with their own unique use cases and strengths, but they also each have their shortcomings and dependencies—as I'm sure you will have realized by now, just from reading the earlier chapters.

As AI continues to evolve and become more integrated into our lives, we need to understand what this means for us as individuals. How will AI impact our careers? How will it shape our education? How will it change our daily lives? What are our rights in an AI-driven world? And what is our role in shaping this new world with AI?

In this chapter, we'll delve into these questions, providing insights and guidance on the future of AI and its implications for you. We'll discuss the potential impact of AI on various careers, the role of education in preparing for an AI-driven future, the potential impact of AI on daily life, the ethical and legal implications of AI, and the role of individuals in shaping the future of AI. Let's begin this journey by exploring the potential impact of AI on your career in the next section.

AI and Your Career

The rise of AI is set to have profound implications for the world of work. From automation and augmentation to creating new roles and industries, AI is significantly reshaping the career landscape.

One of the most discussed aspects of AI's impact on careers is automation. AI has the potential to automate a wide range of tasks, from routine and repetitive tasks to more complex and cognitive tasks. This could lead to significant changes in many jobs and even make some jobs obsolete.

Klarna case study

An example of this is the Swedish fintech company Klarna, an online lending platform integrated as a native purchasing option on many major websites. They launched an AI assistant in 2024 and it immediately got international headlines for doing the equivalent work of 700 full-time customer service agents. In total, Klarna employed about 4000 people at the time, none of which were customer service agents, as they outsource that entire aspect of their business to others.

Klarna made this information public and immediately faced backlash from countless outlets like CBS News, Forbes, and many others. The company stated that they made the information public to inform policymakers that this shift in the job market isn't something that is coming, it is something that is here now. Klarna urges society to start thinking about how we want to navigate this crisis. Klarna made it clear that in the short term, they had no intentions of laying off people as a result of the customer service AI chatbots, but since they already had that function outsourced, it's not necessarily very comforting. In addition Klarna reduced their employee count by about 800 people in 2022, none of which were customer service agents either, and that they have stopped hiring all together since late 2023. Klarna has also stated publicly that they are shrinking their company not by layoffs but by "natural attrition." Instead, Klarna tries to apply AI across all products and services to keep growing the company's financials without growing its employee count.

I am not going to comment on the ethics of what Klarna is doing. I admire their transparency and their realistic approach to implementing this technology. I believe they are choosing to be a front-runner in a way that will attract negative press. But I don't believe they will be alone in this endeavor for very long. And I don't believe they are doing anything wrong. They are keeping up with technology to stay competitive, and on top of that, they are publicly showing how to do it responsibly.

Klarna says their customer service bots are handling two-thirds of their customer service inquiries and that it is on par with humans in regards to customer satisfaction levels, according to customer surveys. They also boast that it lowered repeat inquiries from the same customers with roughly 25%. I believe it's safe to say that quality of work has not decreased as a result of this "optimization," which turns them into a perfect case study—one that others will attempt to replicate.

However, while AI will undoubtedly automate certain tasks, it's important to remember that automation does not necessarily mean job elimination. In many cases, AI will augment human work rather than replace it. By taking over routine tasks, AI can free up time for workers to focus on more complex, creative, and strategic tasks.

Moreover, while AI may automate certain jobs, it will also create new ones. These new jobs may involve developing, managing, or working with AI systems. They may also involve roles that we can't even imagine yet, just as the rise of the Internet created unimaginable roles a few decades ago.

To prepare for these changes, it's important to develop skills that will be in demand in an AI-driven world. These include technical skills, such as data analysis and machine learning, as well as soft skills, such as creativity, critical thinking, and emotional intelligence. It's also important to embrace lifelong learning, as the rapid pace of AI development means that the skills needed for success are constantly evolving.

AI and Your Education

Education plays a crucial role in preparing for an AI-driven future. As AI reshapes the career landscape, it's important to acquire the knowledge and skills that will be in demand. This involves not just formal education but also lifelong learning and continuous skill development.

In terms of formal education, there's a growing demand for skills in AI and related fields like data science, machine learning, and robotics. Many universities and colleges now offer courses and programs in these areas, and there are also many online platforms that offer courses in AI and related fields.

However, more than technical skills will be in demand in an AI-driven world. Soft skills, such as creativity, critical thinking, emotional intelligence, and interpersonal skills, will also be crucial. These are skills that are difficult for AI to replicate and that can complement AI capabilities.

Moreover, as AI becomes more integrated into various fields, there will be a growing need for professionals who understand both their field and AI. For example, in health care, there will be a need for professionals who understand both medicine and AI, and in law, there will be a need for professionals who understand both legal principles and AI.

Lifelong learning is also crucial in an AI-driven world. The rapid pace of AI development means that the knowledge and skills needed for success are constantly evolving. To keep up with these changes, it's important to embrace a mindset of continuous learning and to take advantage of the many resources available for learning about AI.

AI and Your Daily Life

The impact of AI on our daily lives is becoming increasingly evident. From the way we communicate and consume information to how we make decisions and interact with the world around us, AI is reshaping our everyday experiences in significant ways.

One of the most visible ways AI impacts our daily lives is through personalization. AI algorithms power the personalized recommendations we receive on platforms like Netflix, Spotify, and Amazon. These algorithms analyze our behavior and preferences to recommend content, products, or services that we might like.

AI also plays a significant role in our communications. Features like predictive text, voice recognition, and translation services are all powered by AI. These tools can help us communicate more efficiently and effectively and can even enable new forms of communication.

AI is also changing the way we make decisions. From navigation apps that help us find the best route to our destination, to financial apps that help us manage our money, to health apps that help us track and improve our health, AI is increasingly helping us make informed decisions based on data.

However, while these applications of AI can bring many benefits, they also raise important questions and challenges. For example, how do we ensure that AI respects our privacy and doesn't misuse our data? How do we ensure that AI recommendations don't limit our exposure to diverse perspectives and experiences? And how do we ensure that we maintain a sense of agency and control in a world increasingly shaped by AI?

AI and Your Rights

As AI becomes more integrated into our lives, it raises important ethical and legal questions. These questions relate to our rights as individuals, including our right to privacy, our right to fair treatment, and our right to control our own data and decisions.

One of the most pressing issues in this area is data privacy. AI systems often rely on large amounts of data to function, and this data often includes personal information. How do we ensure that this data is used responsibly and that our privacy is respected? How do we navigate the trade-off between the benefits of personalization and the risks of privacy invasion?

Another critical issue is algorithmic fairness. AI systems can inadvertently perpetuate or even amplify existing biases, leading to unfair outcomes. How do we ensure that AI systems treat all individuals fairly, regardless of their race, gender, age, or other characteristics? How do we ensure transparency and accountability in AI decision-making?

A related issue is the control over our own data and decisions. As AI systems make more decisions on our behalf, how do we ensure that we maintain a sense of agency and control? How do we ensure that we have the ability to understand, question, and override AI decisions?

Addressing these issues requires a combination of technological solutions, legal frameworks, and ethical guidelines. It also requires public engagement and dialogue to ensure that the development and use of AI align with our values and aspirations.

AI and Your Role in Society

As AI continues to advance and permeate various aspects of society, it's essential to consider the broader societal implications and our role as individuals in shaping these outcomes. This involves understanding the potential societal impacts of AI, engaging in dialogue and decision-making about AI, and advocating for responsible and equitable AI use.

AI has the potential to bring about significant societal benefits, such as increased productivity, improved health care, and enhanced accessibility. However, it also raises important societal challenges, such as job displacement, privacy concerns, and the concentration of power in the hands of a few tech companies.

As individuals, we have a role to play in shaping the societal impact of AI. This involves staying informed about AI and its implications, participating in public dialogue and decision-making about AI, and advocating for responsible and equitable AI use.

Public dialogue and decision-making about AI can take various forms, from public consultations and citizen juries to online discussions and social media campaigns. These processes help ensure that the development and use of AI align with societal values and aspirations.

Advocacy for responsible and equitable AI use can involve various activities, from promoting ethical AI practices in your workplace to supporting organizations that work on AI ethics to advocating for laws and regulations that promote responsible AI use.

Preparing for an AI-Driven Future

As we look toward an AI-driven future, it's essential to consider how we can prepare ourselves to navigate this new landscape effectively. This involves learning about AI, developing relevant skills, and engaging with AI in a responsible and informed way.

Learning about AI is a crucial first step. This involves understanding what AI is, how it works, and what its capabilities and limitations are. There are many resources available for learning about AI, from online courses and tutorials to books and articles, to podcasts and videos. These resources can help you build a solid foundation of AI knowledge.

Developing relevant skills is another crucial aspect of preparing for an AI-driven future. This includes technical skills, such as programming and data analysis, as well as soft skills, such as critical thinking and ethical reasoning. It also includes the ability to interact effectively with AI systems and understand how to use them effectively and responsibly.

Engaging with AI in a responsible and informed way is also crucial. This involves understanding the ethical and societal implications of AI, advocating for responsible AI use, and participating in public dialogue and decision-making about AI. It also involves understanding your rights in an AI-driven world, such as your right to data privacy and fair treatment.

In conclusion, preparing for an AI-driven future is a multifaceted process that involves learning, skill development, and active engagement. By taking these steps, you can equip yourself to navigate the AI landscape effectively and play a role in shaping the future of AI.

Conclusion

The future of AI is a journey that we are all part of. By understanding AI, developing relevant skills, and engaging with AI in a responsible and informed way, we can navigate this journey effectively and play a role in shaping a future that aligns with our values and aspirations. The journey toward that future starts with understanding and engaging with AI today.

Understanding AI, its capabilities, and its limitations is a crucial part of navigating an AI-driven world. This understanding can help guide your interactions with AI, your career choices, and your educational pursuits.

CHAPTER 9

The AI Gold Rush and the Future of Business

The advent of AI has sparked what many are calling a new gold rush. Just as the original gold rush of the 19th century brought prospectors, entrepreneurs, and pioneers to uncharted territories in search of fortune, the AI gold rush is attracting a new generation of innovators and business leaders eager to capitalize on the transformative potential of AI.

The AI gold rush is characterized by a surge of activity and investment in AI technologies. Businesses across industries are exploring ways to integrate AI into their operations, products, and services. Startups are emerging with innovative AI solutions, and investors are pouring billions of dollars into AI research and development.

However, just like the original gold rush, the AI gold rush presents both opportunities and challenges. On the one hand, AI has the potential to revolutionize industries, create new business models, and generate significant economic value. On the other hand, navigating the AI landscape requires technical expertise, strategic foresight, and ethical consideration.

I like the comparison between the two eras, especially because the real winners of the original gold rush weren't the prospectors, it was the people selling the land lots and shovels. The equivalent of that today are

© Jonas Bjerg 2024
J. Bjerg, *The Early-Career Professional's Guide to Generative AI*,
https://doi.org/10.1007/979-8-8688-0456-4_9

companies like Nvidia who sell the compute-power (computer chips). There are open source alternatives starting out, but the head start Nvidia has almost constitutes a monopoly. As of 2024, Nvidia has 80% of the global GPU market, AMD has 19%, and Intel has 1%. According to their own income statements, Nvidia's net income is expected to be US$ 29 billion in 2024, that's up more than 7x what they made the previous year— US$ 4 billion. I believe that's enough market dominance and enough money earned in one year to constitute never needing to sell anything other than shovels for the foreseeable future.

In this chapter, we will delve into the AI gold rush and its implications for businesses. We will explore the new business opportunities that AI is creating, the disruptive impact of AI on various industries, the importance of having a strategic approach to AI, the ethical considerations for businesses in the AI era, and the skills that will be in demand in the AI era.

As we embark on this exploration, it's important to remember that the AI gold rush is not just about technology. It's also about people, businesses, and society. It's about how we harness the power of AI to create value, drive innovation, and shape a better future. Let's begin this journey by exploring the new business opportunities that AI is creating in the next section.

New Business Opportunities

The AI gold rush is creating a wealth of new business opportunities. From AI services and tools to AI platforms, businesses across industries are finding innovative ways to leverage AI to create value.

One of the most significant opportunities lies in AI services. These are services that leverage AI to provide value to customers. Examples include AI-powered recommendation systems, customer service chatbots, and predictive analytics services. These services can help businesses improve customer experience, optimize operations, and make more informed decisions.

Another significant opportunity lies in AI tools. These are tools that enable businesses to develop, manage, and use AI systems. Examples include machine learning platforms, data visualization tools, and AI development kits. These tools can help businesses accelerate their AI development, improve their AI capabilities, and reduce their AI costs.

A third significant opportunity lies in AI platforms. These are platforms that provide a suite of AI services and tools, often combined with cloud computing resources. Examples include Google's AI Platform, Amazon's AWS AI, and Microsoft's Azure AI. These platforms can help businesses access state-of-the-art AI technologies, scale their AI operations, and innovate faster.

However, capitalizing on these opportunities requires more than just technical expertise. It also requires a deep understanding of customer needs, a strategic approach to AI adoption, and a commitment to ethical AI use. In the next section, we'll discuss the disruptive impact of AI on various industries, exploring how AI is changing business models, operations, and customer experiences.

AI Disruption

The AI gold rush is not just creating new opportunities; it's also disrupting existing industries. By changing the way businesses operate, serve customers, and compete, AI is reshaping the business landscape in significant ways.

One of the most profound impacts of AI is on business models. AI enables new types of business models that were not possible before. For example, AI can enable businesses to offer personalized products and services at scale, create platform-based business models that leverage network effects, and monetize data through AI-powered insights.

AI is also transforming business operations. By automating routine tasks, optimizing processes, and enhancing decision-making, AI can significantly improve operational efficiency and effectiveness. For example, AI can automate data entry tasks, optimize supply chain operations, and enhance financial forecasting.

Furthermore, AI is reshaping customer experiences. By personalizing customer interactions, predicting customer needs, and enhancing customer service, AI can significantly improve customer satisfaction and loyalty. For example, AI can personalize product recommendations, predict customer churn, and enhance customer service through chatbots.

However, while AI can bring significant benefits, it also presents challenges. Businesses need to navigate issues such as data privacy, algorithmic bias, and job displacement. They also need to manage the strategic, organizational, and cultural changes associated with AI adoption.

AI Strategy

As AI continues to reshape the business landscape, having a strategic approach to AI becomes increasingly important. An effective AI strategy can help businesses capitalize on the opportunities of AI, navigate the challenges of AI, and create sustainable competitive advantage.

One key element of an AI strategy is AI integration. This involves integrating AI into various aspects of the business, from operations and products to customer service and decision-making. AI integration requires a deep understanding of the business, a clear vision of how AI can create value, and a robust plan for AI implementation.

Another key element of an AI strategy is AI innovation. This involves using AI to drive innovation, whether by creating new products and services, improving existing ones, or reimagining business processes. AI innovation requires a culture of experimentation, a commitment to learning, and a willingness to take risks.

A third key element of an AI strategy is AI governance. This involves establishing policies and procedures for the ethical and responsible use of AI. AI governance requires a deep understanding of the ethical and societal implications of AI, a commitment to transparency and accountability, and a robust framework for AI ethics and compliance.

Developing an effective AI strategy is not a one-time effort; it's an ongoing process that requires continuous learning, adaptation, and improvement. It requires a deep understanding of AI and its implications, a clear vision of how AI can create value, and a strong commitment to ethical and responsible AI use.

AI Ethics

As businesses navigate the AI gold rush, ethical considerations become increasingly important. From data privacy and algorithmic fairness to AI transparency and accountability, businesses need to address a range of ethical issues to ensure responsible and trustworthy AI use.

Data privacy is a key ethical issue in the AI era. As AI systems often rely on large amounts of data, including personal data, businesses need to ensure that they handle this data responsibly. This involves obtaining informed consent for data collection, ensuring data security, and respecting individuals' rights to access, correct, and delete their data.

Algorithmic fairness is another key ethical issue. As AI systems can inadvertently perpetuate or amplify biases, businesses need to ensure that their AI systems treat individuals fairly. This involves assessing and mitigating biases in data, algorithms, and decision-making processes and ensuring that AI systems do not discriminate against individuals based on their race, gender, age, or other characteristics.

AI transparency and accountability are also crucial. As AI systems often make decisions that affect individuals, businesses need to ensure that these decisions are transparent and accountable. This involves explaining how AI systems make decisions, providing mechanisms for individuals to question and challenge AI decisions, and taking responsibility for the impacts of AI decisions.

Addressing these ethical issues requires a combination of technological solutions, ethical guidelines, and legal frameworks. It also requires a commitment to ethical principles, a culture of ethical awareness, and a robust system of ethics governance.

AI Skills

As the AI gold rush continues, the demand for AI skills is growing rapidly. These skills are not just technical; they also include soft skills and AI literacy, which are crucial for navigating the AI landscape effectively.

Technical skills are the foundation of AI expertise. These include skills in areas like programming, data analysis, machine learning, and AI system design. These skills are crucial for developing, managing, and optimizing AI systems. They are in high demand across industries, and they can provide a significant competitive advantage in the AI era.

Soft skills, while often overlooked, are equally important in the AI era. These include skills like problem-solving, critical thinking, creativity, and communication. These skills are crucial for understanding and addressing the complex challenges of AI, from ethical dilemmas to strategic decisions. They are also crucial for working effectively with AI systems, which often require human oversight, guidance, and interpretation.

AI literacy is another crucial skill in the AI era. This involves understanding what AI is, how it works, and what its capabilities and limitations are. It also involves understanding the ethical and societal implications of AI and knowing how to use AI responsibly and effectively. AI literacy is important not just for AI professionals but for everyone, as AI becomes increasingly integrated into our lives and work.

Developing these skills requires a commitment to learning and development, a culture of curiosity and exploration, and a range of learning resources, from formal education and training to self-directed learning and experiential learning.

Preparing Your Business for the AI Era

As the AI gold rush accelerates, businesses need to prepare themselves to navigate the AI landscape effectively. This involves adopting AI strategically, developing AI talent, and managing AI risks.

Adopting AI strategically is a crucial first step. This involves identifying where and how AI can create value in your business, developing a clear plan for AI adoption, and integrating AI into your business operations, products, and services. It also involves innovating with AI, using AI to create new products, services, and business models.

Developing AI talent is another crucial step. This involves not just hiring AI experts but also developing AI skills and literacy across your organization. It involves providing training and development opportunities, fostering a culture of learning and innovation, and creating a work environment that attracts and retains AI talent.

Managing AI risks is also crucial. This involves understanding and mitigating the risks associated with AI, from data privacy and security risks to ethical and reputational risks. It involves establishing robust policies and procedures for AI ethics and compliance and ensuring that you have the right governance structures in place to oversee your AI activities.

Preparing your business for the AI era is not a one-time effort; it's an ongoing process that requires continuous learning, adaptation, and improvement. It requires a deep understanding of AI and its implications, a clear vision of how AI can create value, and a strong commitment to ethical and responsible AI use.

Conclusion

The AI gold rush is a journey that we are all part of and it is reshaping the business landscape, creating new opportunities, disrupting existing industries, and raising important ethical and societal questions. As businesses navigate this landscape, they need to understand AI, adopt AI strategically, develop AI talent, and manage AI risks.

By understanding AI, adopting AI strategically, developing AI talent, and managing AI risks, businesses can navigate this journey effectively and play a role in shaping the future of AI. The journey toward that future starts with understanding and engaging with AI today.

CHAPTER 10

Tips and Tricks for Prompt Engineering

Artificial intelligence, specifically in the realm of Natural Language Processing, has made significant strides in recent years. One of the most notable advancements is OpenAI's language model, GPT-4, and its more accessible variant, ChatGPT. Which is why this model will be used as the playground for this "tips and tricks"-guide. Even though there are other models out there, these tricks should work on all of them, as all these types of language models each function in a similar way. They all have the ability to generate human-like text, making them incredibly useful for a wide range of applications, from drafting emails to writing code and even authoring articles.

However, to harness the full power of ChatGPT, one must master the art of "prompt engineering." This term refers to the practice of crafting effective prompts to guide the AI in generating the desired output. A well-engineered prompt can mean the difference between a response that is vague and off-topic and one that is precise, insightful, and contextually appropriate.

© Jonas Bjerg 2024
J. Bjerg, *The Early-Career Professional's Guide to Generative AI*,
https://doi.org/10.1007/979-8-8688-0456-4_10

Prompt engineering is not just about asking the right questions; it's about providing the right context, setting the right tone, and giving the right instructions. It's about understanding the nuances of language and the intricacies of the AI model. It's about creativity, experimentation, and continuous learning.

In this chapter, we will delve into the world of prompt engineering with ChatGPT. We will explore the basic workings of ChatGPT, the art of crafting prompts, the use of instructions to guide responses, the role of context, and advanced prompt engineering techniques. We will also provide practical examples and exercises to help you hone your prompt engineering skills.

Whether you're a developer looking to integrate ChatGPT into your application, a business professional using ChatGPT to automate tasks, or a curious individual exploring the possibilities of AI, this chapter will provide you with the knowledge and skills to use ChatGPT effectively.

So, let's embark on this journey into the world of prompt engineering, starting with an understanding of the basics of ChatGPT in the next section.

Understanding the Basics of ChatGPT

ChatGPT, a variant of the larger GPT-4 model developed by OpenAI, is a powerful language model that generates human-like text based on the prompts it's given. To effectively use ChatGPT, it's important to understand the basics of how it works and what influences its output.

At its core, ChatGPT is a transformer-based neural network trained on a diverse range of Internet text. It uses a machine learning technique known as unsupervised learning, meaning it learns patterns and structures in the data it was trained on without being explicitly told what to look for. This allows it to generate creative and coherent responses to a wide variety of prompts.

When given a prompt, ChatGPT generates a response word by word. It uses the context of the prompt and the words it has already generated to predict the next word. This prediction is based on a probability distribution over the entire vocabulary, meaning it considers all possible words and selects the one with the highest probability given the context.

Several Factors Influence the Output of ChatGPT

- **The Prompt:** The prompt you provide is the primary factor that influences ChatGPT's response. The more specific and detailed your prompt, the more likely ChatGPT is to generate a relevant and accurate response.

- **The Model's Training Data:** ChatGPT's responses are influenced by the data it was trained on. It can generate text on a wide range of topics and in a variety of styles because it was trained on a diverse range of Internet text.

- **The Temperature Parameter:** The temperature parameter controls the randomness of ChatGPT's responses. A higher temperature leads to more random responses, while a lower temperature leads to more deterministic responses.

- **The Max Tokens Parameter:** The max tokens parameter controls the length of ChatGPT's responses. It sets a limit on the number of tokens (words or parts of words) that ChatGPT can generate.

Understanding these basics of ChatGPT is the first step in mastering prompt engineering. With this foundation, you can start to explore the art of crafting effective prompts, which we'll discuss in the next section.

The Art of Crafting Prompts

Crafting effective prompts is a crucial aspect of using ChatGPT to its full potential. The right prompt can guide the model to generate insightful, relevant, and creative responses. Here, we'll explore strategies for crafting prompts that are clear, specific, and contextually rich.

- **Clarity:** Clarity is key in prompt engineering. Remember, while ChatGPT is a sophisticated model, it doesn't possess human-like understanding. It generates responses based on patterns it learned during training. Therefore, it's important to make your prompts as clear as possible. Avoid ambiguity and ensure your prompt accurately represents what you want the model to generate. For instance, instead of asking, "Can you tell me about it?" which is vague, you might ask, "Can you provide a summary of the novel '1984' by George Orwell?"

- **Specificity:** Being specific in your prompts helps guide ChatGPT toward the desired output. If you want a certain style, tone, or format, specify it in your prompt. For example, if you're looking for a brief, formal summary of a topic, you might start your prompt with "In a concise and formal manner, summarize..." If you want a response in a specific format, like a poem or a script, indicate this in your prompt.

- **Context:** Providing context in your prompts helps ChatGPT generate more relevant and coherent responses. This is especially important for longer conversations where context from previous exchanges is crucial for continuity. For instance, if you're asking ChatGPT to continue a story, make sure to include the relevant parts of the story in your prompt.

Here's an example of a well-crafted prompt that uses these strategies: "Imagine you're a historian writing in a formal and scholarly tone. Provide a detailed analysis of the impact of the Industrial Revolution on 19th-century European society."

In this prompt, the role of the historian provides context, the request for a detailed analysis ensures specificity, and the formal, scholarly tone adds clarity about the style of the desired response.

Mastering the art of crafting prompts takes practice and experimentation. Feel free to iterate on your prompts and try different approaches. In the next section, we'll delve deeper into how to use instructions within prompts to guide ChatGPT's responses.

Using Instructions to Guide Responses

Incorporating instructions within your prompts is a powerful way to guide ChatGPT's responses. By specifying the format, tone, and content you desire, you can significantly influence the output. Let's explore these techniques in more detail.

- **Specifying Format:** If you need the response in a particular format, make sure to specify this in your prompt. For instance, if you want a list of bullet points, you could start your prompt with "Provide the following information in the form of bullet points." If you want a dialogue, you could set up your prompt as a script, with character names followed by colons.

- **Setting the Tone:** The tone of the response can be guided by the language used in the prompt. If you want a formal tone, use formal language in your prompt. If you want a casual tone, use casual language. You can also explicitly state the desired tone, such as "In a casual and conversational tone, explain..."

- **Guiding the Content:** You can guide the content of the response by being specific about what you want. If you want a summary, specify that you want a summary. If you want a detailed analysis, ask for a detailed analysis. You can also ask ChatGPT to think step by step or debate the pros and cons before settling on an answer.

Here's an example of a prompt with clear instructions: "In a formal and scholarly tone, provide a detailed analysis of Shakespeare's use of iambic pentameter in his sonnets. Your analysis should include

- An explanation of what iambic pentameter is.

- Examples of its use in Shakespeare's sonnets.

- Discussion of how it contributes to the overall effect of the sonnets."

In this prompt, the instructions are clear about the format (a detailed analysis), the tone (formal and scholarly), and the content (iambic pentameter in Shakespeare's sonnets).

Remember, while these techniques can guide ChatGPT's responses, they only guarantee a perfect response sometimes. ChatGPT is a probabilistic model and can sometimes produce unexpected results. Experiment with different instructions and learn from the outcomes to improve your prompts over time.

The Role of Context in Prompt Engineering

Context plays a pivotal role in prompt engineering. It helps ChatGPT understand the situation, maintain continuity in a conversation, and generate relevant responses. Here's how you can effectively use context in your prompts.

- **Providing Background Information:** When you're asking ChatGPT to generate content on a specific topic, it can be helpful to provide some background information in your prompt. This sets the stage for the AI and helps it generate more relevant and accurate responses. For example, if you're asking for a summary of a specific event in World War II, start your prompt with a brief overview of the event.

- **Maintaining Conversation Continuity:** In a conversation or a series of related prompts, it's important to include the relevant parts of the previous exchanges in your new prompt. This helps ChatGPT maintain continuity and context. For instance, if you're asking ChatGPT to continue a story or a conversation, make sure to include the last few exchanges or the relevant parts of the story in your prompt.

- **Guiding the AI's Persona:** If you want ChatGPT to take on a specific persona (like a historian, a scientist, or a fictional character), you can set this context in your prompt. For example, you might start your prompt with "As a historian, explain..." This context helps guide how ChatGPT generates its responses.

Here's an example of a prompt that effectively uses context: "We've been discussing the causes of the French Revolution. As a historian, provide a detailed analysis of the role of economic factors in sparking the revolution."

In this prompt, the first sentence provides context about the ongoing conversation, and the instruction to the AI to take on the persona of a historian provides additional context that guides the response.

Remember, while providing context is important, ChatGPT has a limit to how much text it can consider at once (known as the "context window"). This window is increasing with each release. In September 2021, the limit was 2048 tokens for GPT-3, which includes both the prompt and the response. The later models have much larger capabilities: GPT-35-turbo has a limit of 4096 tokens, GPT-4 has a limit of 8192 tokens and GPT-4-32k has a limit of 32,769 tokens.

Advanced Prompt Engineering Techniques

Once you've mastered the basics of prompt engineering, you can start to explore more advanced techniques. These techniques can help you handle complex tasks, manage errors, and optimize the performance of ChatGPT. Let's delve into these strategies.

- **Handling Complex Tasks:** For complex tasks, it can be helpful to break down the task into smaller parts and guide ChatGPT through each part step by step. For instance, if you're asking ChatGPT to write an essay, you might first ask it to generate an outline, then ask it to expand each point of the outline into a paragraph.

- **Managing Errors:** Despite its impressive capabilities, ChatGPT can sometimes make errors or generate outputs that don't meet your expectations. In such cases, you can use your prompts to correct the errors and guide the model toward the desired output. For instance, if ChatGPT generates a factually incorrect response, you can point out the error in your next prompt and ask the model to correct it.

- **Optimizing Performance:** There are several ways to optimize the performance of ChatGPT. One way is to experiment with the model's parameters, such as the temperature and max tokens settings. Another way is to iterate on your prompts, trying different phrasings and instructions to see what produces the best results. You can also use reinforcement learning from human feedback, a technique used by OpenAI to fine-tune ChatGPT to train the model on specific tasks.

Here's an example of an advanced prompt: "We're working on an essay about climate change. Let's start by generating an outline. The essay should have an introduction, three main points, each with supporting evidence, and a conclusion."

In this prompt, a complex task (writing an essay) is broken down into a simpler task (generating an outline), providing a clear structure for ChatGPT to follow.

Remember, advanced prompt engineering involves much experimentation and learning from trial and error. Keep going even if your initial attempts don't produce the desired results. Keep iterating on your prompts, learning from the outcomes, and refining your techniques.

Practical Examples and Exercises

In this section, we'll provide practical examples and exercises to help you apply and practice the prompt engineering techniques discussed in this chapter. These exercises are designed to cover a range of scenarios and challenges, giving you a comprehensive practice experience.

Exercise 1: Crafting Clear and Specific Prompts

Task: Ask ChatGPT to generate a brief, formal report on the impact of climate change on polar bears.

Exercise 2: Using Instructions to Guide Responses

Task: Ask ChatGPT to write a dialogue between two characters, Alice and Bob, discussing their plans for a weekend hiking trip. The conversation should be casual and conversational.

Exercise 3: Providing Context

Task: Continue the following story: "In a bustling city filled with towering skyscrapers and neon lights, a young detective named Sam was on the trail of a notorious art thief."

Exercise 4: Handling Complex Tasks

Task: Ask ChatGPT to write an essay on the benefits and challenges of renewable energy. Start by asking it to generate an outline, then ask it to expand each point of the outline into a paragraph.

Exercise 5: Managing Errors

Task: If ChatGPT generates a factually incorrect response or a response that doesn't meet your expectations, use your prompt to correct the error and guide the model toward the desired output.

Exercise 6: Optimizing Performance

Task: Experiment with different phrasings, instructions, and model parameters to optimize the performance of ChatGPT. Try to generate the best possible response to a prompt of your choice.

Remember, these exercises are just a starting point. Feel free to create your own exercises based on your needs and interests. The key is to practice, experiment, and learn from the outcomes. With time and practice, you'll become more proficient at prompt engineering and be able to harness the full power of ChatGPT.

Conclusion

In this chapter, we've explored how ChatGPT works and the factors that influence its output. We've delved into the art of crafting prompts and the role of context in prompt engineering. We've also discussed advanced techniques for handling complex tasks, managing errors, and optimizing performance.

Remember, prompt engineering is as much an art as a science. It involves creativity, experimentation, and continuous learning. Feel free to try different approaches, learn from the outcomes, and refine your techniques.

As we move forward into an era where AI plays an increasingly prominent role in our lives, effectively communicating with these systems will become an invaluable skill. So keep practicing, learning, and exploring the fascinating world of prompt engineering with ChatGPT or any of the other alternatives out there—Anthropic's Claude is pretty impressive in its own right.

CHAPTER 11

Ethical Implications and Societal Impact of AI

The rapid progression of AI technology raises several ethical concerns. A key issue is privacy. AI systems need large amounts of data for learning and improvement, which can lead to a dilemma between beneficial data collection and invasion of personal privacy. For instance, AI-powered surveillance systems, though beneficial for security, often stir debates about the extent to which personal data is being monitored and stored.

Data security is another major ethical issue. AI systems often handle sensitive data, making them targets for cyberattacks and potential data breaches. This situation poses questions about the resilience of AI systems against such threats and the steps taken to protect sensitive information.

Another significant ethical challenge is bias in AI algorithms. AI learns from data, which may have inherent biases, causing these systems to continue or worsen these biases. There have been cases where AI algorithms in recruitment, law enforcement, and loan approvals have exhibited biases against certain groups, leading to unfair and discriminatory results.

© Jonas Bjerg 2024
J. Bjerg, *The Early-Career Professional's Guide to Generative AI*,
https://doi.org/10.1007/979-8-8688-0456-4_11

These instances highlight the importance of cautious consideration and proactive actions to manage the ethical implications of AI. This ensures that the development of AI technology benefits society while minimizing potential harm.

Societal Impact

AI is profoundly transforming various facets of society, from how we work and learn to how we conduct our daily lives.

Employment and Job Market Dynamics

AI is revolutionizing the employment landscape. On the one hand, it's creating new job opportunities, particularly in fields related to technology, data analysis, and AI development. There's a growing demand for professionals skilled in these areas, which includes roles like AI researchers, data scientists, and machine learning engineers. On the other hand, AI is also leading to the displacement of jobs, particularly those involving routine, repetitive tasks. Jobs in manufacturing, data entry, and even some aspects of customer service are increasingly automated. This shift is creating a need for a more adaptable, technically proficient workforce committed to lifelong learning. The challenge lies in preparing current and future workers for this change, necessitating reevaluating and revamping educational and training programs to align with these evolving needs.

Transformation in Education

In the realm of education, AI is enabling more personalized and effective learning experiences. AI-driven educational tools and platforms can tailor content to meet individual student needs, recognizing their unique strengths, weaknesses, and learning styles. This personalization can potentially lead to better student engagement and improved educational outcomes. AI can also assist teachers by automating administrative tasks, allowing them more time to focus on teaching and student interaction. However, this shift also brings challenges, such as ensuring equitable access to these advanced technologies for all students and addressing privacy concerns related to student data.

Impact on Daily Life

AI's integration into daily life is becoming more pronounced. Smart home technologies, like AI-powered thermostats and assistants, are changing how we interact with our living spaces, making them more efficient and responsive to our needs. In health care, AI is being used for diagnostic purposes, personalized treatment plans, and even in robotic surgeries, offering the potential for more accurate and efficient care. The entertainment industry, too, is leveraging AI for content recommendation, special effects, and even the creation of music and art. However, this pervasive integration raises concerns about over-reliance on technology, potential loss of privacy, and diminishing certain human skills and interactions. For instance, the ease of AI-driven solutions can decrease critical thinking and problem-solving skills as tasks become more automated.

The Need for a Balanced Approach

The dual impact of AI—its immense benefits and potential drawbacks—underscores the need for a balanced approach to technology adoption. In many cases, embracing AI's advancements is crucial, enhancing efficiency, convenience, and even safety. However, it is equally important to be mindful of the potential negative effects, such as job displacement, privacy issues, and over-reliance on technology. This balance can be achieved through thoughtful policy-making, ethical AI development guidelines, and a focus on education and training that prepare individuals for a world where AI plays a central role. Ensuring that AI benefits society while minimizing its adverse effects will be a key challenge and responsibility for technology creators and policymakers.

AI in Governance and Policy-Making

Integrating AI in governance and policy-making is a rapidly developing field with significant societal implications. AI's ability to process vast amounts of data and identify patterns can enhance decision-making in various governmental functions. However, its application in this domain also raises critical concerns regarding ethics, transparency, and accountability.

In Law Enforcement

AI's role in law enforcement, particularly through predictive policing, can potentially revolutionize public safety. By analyzing data on past crimes, AI systems can predict where and when future crimes might occur, enabling law enforcement agencies to allocate resources more effectively. This proactive approach can potentially reduce crime rates and enhance community safety.

However, using AI in law enforcement also raises significant privacy and bias concerns. There is a risk that these systems may perpetuate existing biases present in the historical data they are trained on, such as racial profiling. Moreover, the widespread surveillance required for data collection can lead to invasions of privacy, raising ethical questions about the extent to which this technology should be employed.

In Judicial Decisions

In the judiciary, AI can assist in data analysis and pattern recognition, leading to increased efficiency and potentially reducing human error in legal decisions. AI tools can help sift through large volumes of case law and legal documents to assist judges and lawyers in their work.

However, this assistance comes with the risk of opaque decision-making processes. Suppose AI systems are used to make or suggest judicial decisions. In that case, it may be challenging to understand the basis of these decisions, as AI algorithms can be complex and not easily interpretable. This lack of transparency can lead to concerns about accountability and fairness in the judicial process.

In Policy Formulation

AI can significantly aid policy formulation by simulating the outcomes of various policy choices. Governments can use AI models to predict different policies' economic, social, and environmental impacts, leading to more informed decision-making.

However, these models are often reliant on historical data, which may not always accurately capture complex human behaviors and societal nuances. Policies based exclusively on AI predictions might overlook critical human factors, leading to unintended consequences. Therefore, while AI can be a valuable tool in policy-making, it should be used in conjunction with human judgment and expertise.

In conclusion, while the integration of AI in governance and policy-making offers promising benefits like efficiency and data-driven insights, it is crucial to approach its implementation with caution. Governments and policymakers must consider the ethical implications, ensure transparency in AI-driven decisions, and maintain accountability. Balancing the technological capabilities of AI with these concerns is essential for its responsible and effective use in governance.

Ethical AI Design and Regulation

The design and regulation of AI with ethical considerations is a complex and crucial task aimed at ensuring that the development and deployment of AI systems are beneficial to society and do not cause unintended harm. Ethical AI design and regulation involve several key principles and approaches across different regions and contexts.

Ethical AI Design's Core Principles

- **Fairness:** AI systems should be designed to be unbiased and non-discriminatory. This involves addressing and mitigating biases in data, algorithms, and their implementation, ensuring that AI decisions do not unjustly favor one group over another.

- **Transparency:** AI systems should be transparent and explainable. Users and affected parties should be able to understand how and why a particular AI decision was made. This transparency is crucial for building trust and accountability.

- **Accountability:** There should be clear mechanisms for holding AI system designers, operators, and deployers accountable. This involves determining responsibility for the decisions made by AI systems and for addressing any negative impacts.

- **Privacy:** AI systems should be designed with respect for user privacy, ensuring that personal data is protected and used ethically.

- **Safety and Security:** AI systems should be secure against tampering and misuse and operate safely under a wide range of conditions.

Global Regulatory Approaches

- **European Union:** The EU is known for its more regulatory approach to AI. Regulations often focus on high-risk AI applications in health care, law enforcement, or transportation. The EU emphasizes the need for transparency, accountability, and human oversight in AI systems, ensuring that AI aids human decision-making rather than replacing it entirely. The EU's approach is often considered stringent but designed to protect citizens' rights and safety.

- **United States:** The United States takes a more industry-led approach to AI regulation. This approach fosters innovation by allowing AI developers more freedom to create and deploy AI technologies. The emphasis is on self-regulation and the development of industry standards and best practices. This method aims to mitigate risks while encouraging technological advancement and economic growth.

- **Other Regions:** Other countries and regions also explore different models of AI regulation, often reflecting their unique cultural, political, and economic contexts. For instance, some countries may prioritize AI's economic and competitive advantages, while others focus more on social welfare and ethical considerations.

The ethical design and regulation of AI are vital for harnessing the benefits of AI technologies while minimizing potential harms. Global approaches to AI regulation reflect efforts to balance technological advancement with ethical responsibility and societal welfare.

As AI continues to evolve, ongoing dialogue and collaboration among governments, industry, academia, and civil society are essential to develop effective and ethical AI governance frameworks.

Public Perception and Education

Public perception and understanding of AI are crucial factors in this technology's societal integration and ethical use. As AI continues to permeate various aspects of life, the general awareness of both its potential benefits and associated risks is becoming increasingly important. Education, in this context, plays a pivotal role in shaping a society that is not only informed about AI advancements but is also prepared to adapt to and interact with these technologies responsibly.

Evolving Public Perception of AI

Public perception of AI has been shaped by a range of factors, from media portrayals to personal experiences with AI-driven technologies. As AI applications become more prevalent in daily life—such as smartphones, home assistants, and online services—the public is gaining a more nuanced understanding of what AI can and cannot do. This growing awareness also brings to light concerns about privacy, job displacement, and ethical use of AI.

The portrayal of AI in popular culture and media often swings between utopian and dystopian narratives, which can influence public perception. While some view AI as a revolutionary technology that will solve many of the world's problems, others fear it as a tool that could lead to significant societal harm. Bridging this gap in understanding is essential for a balanced and realistic perspective on AI.

Role of Education in AI Literacy

Integrating AI literacy into education curricula is a strategic approach to demystify AI and provide a foundational understanding of its principles, capabilities, and limitations. This education should not be limited to technical knowledge but should also encompass ethical, social, and practical aspects of AI. By understanding how AI systems are designed, trained, and applied, individuals can better appreciate the nuances of AI interactions and implications.

AI education can empower individuals to

- **Critically Engage with AI Technologies:** Educated individuals can make informed decisions about how they interact with AI systems, understanding the implications of data sharing, privacy, and security.

- **Advocate for Ethical Use:** With a solid understanding of AI, people can participate in discussions and advocacy related to ethical AI development and deployment, ensuring that these technologies are used for the greater good.

- **Adapt to AI-Driven Changes in the Workforce:** As AI continues to transform the job market, an understanding of AI can help individuals adapt to new roles or modify their skill sets to remain relevant in an evolving job landscape.

Importance of Broad Access to AI Education

To develop a society capable of harnessing AI's benefits while mitigating its risks, it is essential that AI education is accessible to a broad audience. This includes students in formal education settings, lifelong learners, workers in industries impacted by AI, and underserved communities that might be disproportionately affected by AI advancements.

In conclusion, as AI becomes increasingly embedded in our society, public perception and education about AI are more important than ever. A well-informed public that understands AI's capabilities, risks, and ethical implications can contribute to a future where AI is used responsibly and beneficially. Educating a diverse range of individuals to AI literacy is not just a matter of providing technical knowledge. Still, it is also crucial for empowering society to participate actively in shaping how AI impacts our world.

Speculation on the Near Future

The near future of AI presents a spectrum of possibilities that could significantly shape various aspects of society. The next decade will likely witness both optimistic and pessimistic scenarios based on how AI evolves, is regulated, and is integrated into our daily lives.

Optimistic Scenarios

- **Personalized Health care:** AI has the potential to revolutionize health care by enabling more personalized and precise medical treatments. With advancements in AI, we could see more accurate diagnostic tools, tailored treatment plans based on individual genetics and lifestyle, and improved drug discovery processes. AI could also enhance remote patient monitoring and predict disease outbreaks, leading to better preventive health care.

- **Environmental Solutions:** AI could play a crucial role in addressing complex environmental challenges. It can be used to model climate change scenarios, optimize renewable energy systems, monitor deforestation and biodiversity loss, and manage sustainable urban planning. AI's ability to analyze large data sets can help make more informed environmental conservation and resource management decisions.

- **Improved Efficiency in Various Sectors:** AI can improve efficiency across multiple industries. In manufacturing, AI-driven automation can increase production efficiency and reduce waste. In agriculture, AI can help in precision farming, optimizing the use of resources like water and fertilizers. In transportation, AI could lead to safer and more efficient traffic management and the advancement of autonomous vehicles.

Pessimistic Scenarios

- **Job Displacement:** One of the major concerns is that AI could lead to significant job displacement. Automation and AI technologies might replace human roles, particularly in sectors like manufacturing, customer service, and even some professional services. This could lead to increased unemployment and require a major shift in workforce skills.

- **Widening Socioeconomic Gaps:** The unequal distribution of AI benefits could exacerbate socioeconomic inequalities. Countries and individuals with access to advanced AI technologies could surge ahead in terms of economic growth and quality of life, while those without access could lag behind. This digital divide could deepen existing inequalities.

- **Privacy and Ethical Concerns:** As AI systems become more integrated into daily life, there could be significant implications for privacy and ethical considerations. Issues around data collection, surveillance, and the potential misuse of AI technologies could lead to societal concerns and distrust of AI.

Determining Factors

The trajectory of AI shortly will depend on several key factors:

- **Technological Breakthroughs:** AI research and development advancements will shape its capabilities and applications.

- **Regulatory Frameworks:** How governments and international bodies regulate AI will significantly impact its development and deployment. Effective regulation can ensure that AI is used ethically and for the benefit of society.

- **Societal Engagement:** Public perception, understanding, and engagement with AI technologies will influence how these are integrated into society. Education and awareness about AI will play a key role in shaping its role in our future.

Conclusion

The next decade will be pivotal for AI, filled with potential for both positive advancements and challenges. The direction AI takes will hinge on a balance of technological innovation, ethical considerations, regulatory measures, and public engagement. Proactive planning and policies will be crucial in steering AI toward outcomes that benefit society as a whole.

CHAPTER 12

The Platform Shift

In summary, AI's ethical and societal dimensions are intricate and require thoughtful and comprehensive consideration. The challenges presented by AI are diverse and impactful, spanning from individual privacy concerns to broader societal shifts. These challenges, such as privacy issues, data security risks, and the inherent biases in AI algorithms, call for a nuanced and informed approach to AI development and implementation.

The impact of AI on employment and education sectors further illustrates the dual nature of AI's influence. While AI presents opportunities for innovation and efficiency, it also poses risks of job displacement and necessitates a shift in educational paradigms to prepare future generations. The integration of AI literacy into education systems is vital for creating a society that is not only aware of but also equipped to handle the changes and challenges brought about by AI technologies.

Furthermore, the role of AI in governance and policy-making is a testament to its growing influence in decision-making processes. This highlights the critical need for transparency and accountability in AI systems to ensure they are used ethically and responsibly. The varying global regulatory approaches to AI, from the stringent regulations in the European Union to the industry-led guidelines in the United States, reflect the global endeavor to find a balance between encouraging technological innovation and upholding ethical standards.

© Jonas Bjerg 2024
J. Bjerg, *The Early-Career Professional's Guide to Generative AI*,
https://doi.org/10.1007/979-8-8688-0456-4_12

Public perception and education about AI play a pivotal role in determining how society adapts to and interacts with these technologies. Educating the public about AI demystifies the technology and empowers individuals to participate actively in discussions about its ethical use and implications. This education is essential for preparing society for both the optimistic and pessimistic scenarios that the future of AI might present.

As AI advances and becomes more embedded in our daily lives, the importance of these considerations and societal preparedness cannot be overstated. AI's multifaceted implications require a balanced approach that embraces its benefits while conscientiously mitigating its risks. AI's future holds great potential, but realizing this potential in a way that benefits society as a whole will depend on our collective efforts to address these ethical challenges and prepare for its transformative impact.

The genie is out of the bottle now, and there is simply no way to get it back in. A global ban would not slow down the advancement much at this point. But that doesn't mean that advances in AI have to be all bad. Just like with any other technology, there are good and bad actors. I wrote this book to highlight the positives and provide a counterweight to today's doomsday narrative domination of the news. I have hinted at this throughout the book a lot already, but let me state it once and for all plainly here in the final chapter.

I, Jonas Bjerg, don't believe AI will take more jobs than it creates in the foreseeable future. If we take stock of the jobs lost to other technological advancements during the Industrial Revolution and later with automation and computers, the impact on labor paints a much more positive picture. Instead of seeing mass layoffs, I will show in this chapter that the exact opposite has happened every time. With the many AI examples in previous chapters in mind, I hope to have sufficiently made the point that we are far from general AI. Even the latest model from OpenAI, GPT-4o, which can see images and talk on top of reading text, is built by assembling multiple different models and having them seemingly work as one—but they aren't. There is nothing truly "general" about the intelligence in even the most advanced public-facing models I have seen to date. Since its inception, ChatGPT and its competitors have felt very impressive and can look like general intelligence, to an untrained eye, but when asked a simple mathematical question, its shortcomings quickly show. So no, we are far from having a general intelligence that can make us all redundant tomorrow. It will look different 50 years into the future, but for now, I recommend choosing the career you are most interested in and growing with that industry. Instead of trying to outsmart progress, the old advice, "listen to your heart," still holds true in my opinion—when it comes to choosing a profession. But not every profession needs a long education. And not everyone is fortunate enough to be able to choose.

Ten years ago, the industry almost unanimously predicted that blue-collar jobs would be the first to be automated, followed by white-collar and then creative jobs. In many ways the opposite has happened. probably partly because the prediction itself made it easier to get funding to solve the "impossible" creative jobs. Regardless of the reason why, the fact is today the vast majority of AI tools on the market target creative professions. Repetitive blue-collar jobs are next in line. The main industries are transportation and assembly line jobs. I do not doubt that a new assembly-line revolution is close to a reality. Nvidia, Boston Robotics, and Tesla have all unveiled stand-alone assembly worker robotics in recent years. And multiple major car companies have soft released "level 3" autonomy, meaning level 4 and 5's main hurdle soon becomes their illegality on the roads in many countries. So let's look at what actually happened the last time we went through a shift of this scale and what impact it had on jobs in society. We had the industrial revolution and the shift from horses to vehicles happen in the last two centuries

Deep-Dive on Historic Employment Rates in America

We must first collect data to paint the best picture of the many new inventions during the Industrial Revolution and their impact on job employment rates. History books can provide all the noteworthy dates major inventions hit the public market. It is immediately more challenging to find reliable data on employment rates that date back to the 1800s. Fortunately, the census data in America provides a beautiful substitute for just this purpose. Together with the data publicly available from the Organisation for Economic Co-operation and Development (OECD), we can build a relatively reliable timeline of what happened, when, and their impact on employment rates in America. Unfortunately, I haven't found a reliable substitute for the American census data in Europe, so our thought experiment must remain confined to the United States. The only data concern we need to address before combining the two data sources is that the US census calculates its employment rate based on every available person aged above 10, whereas the OECD data only counts people aged above 15. This difference is not a big problem; we only want to see if the upward-moving trend continues throughout both datasets. So, since the two calculation methods vary slightly but stay consistent within each dataset, their employment rates remain comparable.

Interestingly, it also shows the societal shift between then and now. Slavery was still considered normal in the 1800s, so deciding to exclude or include those jobs employed by enslaved people in the 1800s would presumably impact the employment rates much more than a small 5-year age gap. Regardless of how we combine the datasets and rationalize their nature, there is a jump between the two datasets. The sudden change in employment rates in the 1940s and 1950s is probably better explained by the Second World War, which both provided work for many people and significantly impacted the total population numbers, again affecting our employment numbers much more than a 5 year gap.

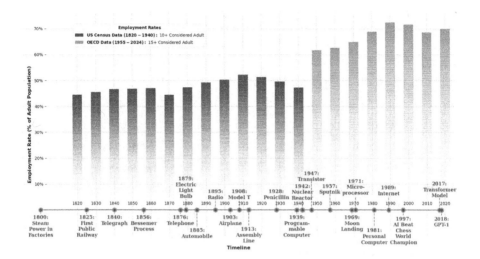

The preceding graph constructs a timeline of when certain major inventions first hit the American market and what the employment rate was in America at the time, represented by the purple and red bars above the timeline. Each color represents which dataset the period is covered by. If the narrative that technological progress leads to job loss, the graph should show a decline in employment rate after the major technological releases. That is, however, not the case. As you can see, the more technological our society has gotten, the higher the employment rate. This shows that as technological advances have improved our industries, the available jobs have not suffered as a result—quite the opposite in fact.

When industries see massive leaps in technological improvement, like the introduction of the automobile in 1885 or the assembly line in 1913, it can seem logical that those technologies would mean we needed fewer people employed to produce the same outcome. Whether it is fewer drivers because a single car can carry more weight or drive further than a horse carriage, or a single worker can increase the overall output in some other way, thanks to technology. When this happens, it is logical to think that this will mean the need for fewer people, but historically, that has not been the case. As the graph shows, as progress has enabled more and more industries, the employment rate has slowly increased decade over decade. In other words, as the achievable outcome increased year over year, the expected output has kept pace. When a new technology enabled an improvement, the supply and demand curves shifted almost immediately, and the industry adapted accordingly—probably by lowering prices. Yes, there will be some bumpy parts during these transition periods. Over the last two centuries, some industries have indeed disappeared overnight. The best example is the Whaling industry. We used to source whale oil for lighting and lubrication, but when petroleum was discovered in the mid-to-late 19th century, the entire industry disappeared.

Another example is the Telegraph, which was shown in the preceding graph, which was introduced in 1840. The telegraph industry was once a vital means of long-distance communication. The entire industry died out instantly when the telephone rose to prominence in 1876. The jobs shifted in both these examples, and the telegrapher's job was replaced with the telephone operator's. Especially in the 1800s and the early parts of the 1900s, it is clear that many of these shifts happened seamlessly, as many of the new jobs didn't require much education or training—this has changed in recent years. This is perhaps the best reason for people to be a little worried about the future, as the biggest current industry in the world is transportation. A job anyone can be trained to do in weeks. All they need is a driver's license. When level 4 or 5 autonomy is publicly available, it is safe to say that those jobs will go the same way as whalers or telegraphers' jobs did. And now, it is hard to say what profession these people can shift to, as few other industries are left where little or no education is required. I am not pessimistic about the future, though; as the telegraphers found other jobs, so will drivers.

Let's now dive a little deeper into this graph and dissect the small decreases in employment rate. The graph clearly shows an overwhelmingly positive growth trajectory, but there are bumps and lows on the way. Is all this simply explained by those shifts?

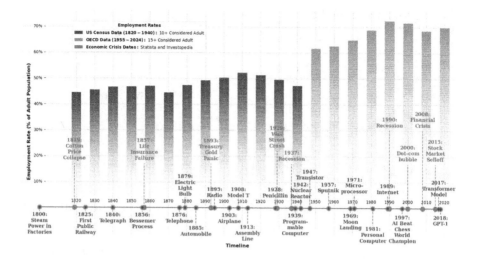

On the preceding graph is introduced an alternative explanation for these bumps in the road. The newly introduced red dots on the timeline represent the different crises and struggles of the financial markets over the last two centuries. These crises far better explain the sudden drop in employment rate, as layoffs often happen immediately after this crisis hits. There can be slight delays, as we see with the life insurance failure in 1857 or the Wall Street crash in 1929, which led to the great depression in the subsequent years. I would gesture that technological improvements are not to blame for these bumps or temporary drops in employment rates and that the financial crises instead bears most of the responsibility for those tough times.

In short, I believe technological advancements have historically helped humanity, and that still holds true today. So don't be afraid of progress. We will still have enough jobs for everyone for many years to come. But yes, when true general intelligence is released, there will have to be many major changes made to our society. Chief among them being to find a new way to distribute wealth in society if everyone can't have a job. But that is a political question and not one we will need to answer in the foreseeable future.

But what will then be the focus in the foreseeable future?

I believe there are three main technological advances coming:

- Personalization at scale
- Integration of Large Language Models in products
- Interface between people and devices

Personalization at Scale

In modern business, artificial intelligence is a key driver of personalization at scale. This concept involves using technology to tailor products, services, and experiences to individual customer preferences and needs, a capability that is becoming increasingly feasible day by day. First movers have already adopted the methodology to create new episodes of their favorite TV shows, currently only in written form, but even that is improving day by day. I have seen examples of entire south park episodes generated entirely with AI in animated video form and everything. As a person who hasn't watched much of the TV show, I must admit I could not tell the difference. Imagine being able to prompt new Harry Potter books and have them feel like they were actually written by J. K. Rowling.

AI facilitates deep learning and data analytics, which allow businesses to parse through vast amounts of data to discern patterns and preferences. With these insights, companies can create highly personalized experiences that not only meet but often anticipate the needs of their customers. This level of personalization is seen in various industries, from retail with customized shopping recommendations to entertainment where streaming services provide tailored playlists and viewing suggestions.

For instance, e-commerce platforms utilize AI to analyze customer browsing and purchase history to offer individualized product suggestions and promotions. This not only enhances the shopping experience but also increases the likelihood of purchases by presenting the most relevant products to each user.

The impact of personalization extends beyond consumer satisfaction to include significant improvements in business performance. By delivering more relevant experiences, businesses see higher engagement rates, increased loyalty, and ultimately, greater revenue. Personalization at scale also enables companies to differentiate themselves in competitive markets, offering unique value propositions that are finely tuned to the demands of their customer base.

Moreover, AI-driven personalization at scale contributes to operational efficiencies. By automating the personalization process, businesses can handle an ever-increasing volume of interactions without corresponding increases in overhead costs. This scalability is crucial for growth, especially in industries where the volume of user data and the speed of transactions are high.

Despite the benefits, challenges such as data privacy concerns, the need for sophisticated data management systems, and the potential for bias in AI algorithms must be addressed. Companies must implement strong data governance policies and invest in robust AI systems that can handle complex data analysis ethically and efficiently.

In conclusion, as AI technology continues to advance, the ability to personalize at scale will become a fundamental aspect of business strategy, driving customer engagement and satisfaction while fostering business growth and innovation.

Integration of Large Language Models in Products

The integration of large language models (LLMs) such as ChatGPT into commercial products represents a transformative shift in how businesses interact with their customers and streamline their operations. These AI-driven tools are not merely technological enhancements; they are reshaping the very fabric of product design and customer engagement.

This capability allows the LLM empowered products to understand and respond to user queries, generate content, and even automate complex tasks that traditionally required human intelligence. By embedding these models into products, companies can offer more intuitive, engaging, and efficient user experiences.

Enhancing User Experience

One of the primary benefits of integrating LLMs into products is the significant enhancement of user experience. These models enable natural language interactions, allowing users to communicate with digital products and services as if they were interacting with a human. This can significantly reduce the learning curve for new users and improve overall satisfaction by making technology more accessible and easier to use.

For example, customer service platforms incorporate LLMs to provide real-time, conversational assistance to users. Instead of navigating through complex menus or waiting in line for a human operator, customers can get immediate help through a chat interface that understands and responds to their queries in natural language.

Innovative Product Features

LLMs also enable the development of innovative product features that were previously unfeasible. For instance, educational technology products can use LLMs to create adaptive learning environments that respond to the needs of individual students, offering personalized feedback and support that enhances the learning experience.

In the creative industries, LLMs are being used to assist with content creation, from drafting articles and reports to generating creative writing and even code. This not only boosts productivity but also allows human creators to focus on higher-level creative processes by automating more routine aspects of content production.

Operational Efficiencies

Beyond enhancing user experience, the integration of LLMs into products can also drive significant operational efficiencies. By automating complex tasks, LLMs can help reduce labor costs and streamline business processes. In industries like legal and finance, where document analysis and data processing are time-consuming, LLMs can quickly analyze large volumes of text, extract relevant information, and even make preliminary decisions based on preset criteria.

Challenges and Considerations

However, integrating LLMs into products comes with its own set of challenges. Issues such as data privacy, the need for large, unbiased training datasets, and the potential for errors in understanding or generating inappropriate content are significant considerations. Moreover, the deployment of LLMs must be managed carefully to complement human workers, not replace them, ensuring that the technology enhances rather than detracts from the human element of service.

In conclusion, the integration of large language models into products is a game-changer for many industries, offering unprecedented opportunities for innovation and efficiency. As these technologies continue to evolve, they will play a crucial role in defining the next generation of digital products and services.

Interface Between People and Devices

The interface between people and devices is undergoing a transformative shift, driven largely by advancements in artificial intelligence. This evolution is not just about enhancing functionality but also about redefining the very nature of interaction between humans and technology. As devices become smarter, they are increasingly capable of anticipating user needs and adapting to user behaviors, offering a more intuitive and seamless experience.

Evolving User Interfaces

Traditional user interfaces, based on buttons and menus, are rapidly being complemented and, in some cases, replaced by more dynamic interfaces. These new interfaces, powered by AI, are capable of understanding and responding to natural human gestures and commands. Voice assistants like Siri, Alexa, and Google Assistant are prime examples of this shift, where verbal commands have largely replaced physical interaction for many functions.

Beyond voice, AI is also enabling interfaces that can interpret visual cues such as facial expressions and body movements, leading to more engaging and human-like interactions. For example, in the gaming and virtual reality industries, AI-driven systems can adjust the game environment in real time based on the player's physical responses, enhancing the immersive experience.

Personalized Interaction

AI enhances the ability of devices to offer personalized interactions. By learning from user data, AI can tailor the functionality and responses of devices to suit individual preferences. In smart homes, for instance, AI can learn residents' routines and adjust lighting, temperature, and even music based on the time of day or the presence of people in the room.

This level of personalization is made possible by the continuous analysis of data collected from user interactions. The devices get smarter over time, improving their predictions and adjustments, which in turn enhances user satisfaction and engagement.

Accessibility and Inclusion

One of the most significant impacts of AI-driven interfaces is the improvement in accessibility and inclusion. AI technologies enable the development of assistive devices that can help individuals with disabilities interact more effectively with technology and their environment. Speech recognition can assist those with visual impairments, while advanced text-to-speech technologies can aid those with hearing impairments.

Furthermore, AI can bridge language barriers, making technology accessible to a broader range of users. Real-time translation services, both spoken and written, are becoming increasingly accurate and instantaneous, enabling people from different linguistic backgrounds to communicate effectively.

Challenges and Ethical Considerations

However, the development of AI-driven interfaces is not without challenges. Privacy concerns are paramount, as these systems often require access to personal data to function optimally. There is also the risk of dependency, where users become overly reliant on AI assistance, potentially diminishing human skills and capabilities.

Additionally, there are ethical considerations in ensuring that these technologies do not exacerbate social inequalities. The design and deployment of AI interfaces must be inclusive, catering to diverse user groups and ensuring that the benefits of technology are accessible to all.

As AI continues to evolve, the interface between people and devices will become increasingly sophisticated, transforming how we interact with technology in our daily lives. This shift promises not only to enhance user experience but also to offer new levels of personalization and accessibility, making technology more intuitive and useful than ever before.

Conclusion: Economic and Societal Impacts of AI

The influence of artificial intelligence extends far beyond individual industries; it has profound economic and societal impacts that are reshaping the global landscape. As noted by industry experts like James

Larmer, Partner at Bain & Company, and Carl Solder, Cisco CTO, in a recent Forbes article by Elise Shaw, "AI is poised to add trillions of dollars to global GDP over the next decade, driving growth through enhanced labor productivity, increased revenue, and greater customer satisfaction."

Economic Growth and Job Creation

James Larmer emphasizes that AI's potential to drive economic growth is significant. By automating routine tasks and optimizing various business processes, AI enables companies to focus resources on more strategic areas, thereby increasing overall productivity. While there is a common concern that AI might displace jobs, the reality is more nuanced. AI is likely to create new job categories even as it renders some obsolete, leading to a net gain in employment opportunities in the long run.

For instance, the adoption of AI in sectors like health care and finance has not only streamlined operations but also created demand for new skill sets, such as AI system management and data analysis. These roles require a new kind of workforce, skilled in both the domain knowledge and the technological aspects of AI.

Business Innovation and Service Enhancement

AI drives innovation by enabling the development of new products and services that were previously unimaginable. Companies are leveraging AI to tap into unmet customer needs, much like Uber revolutionized transportation by combining mobile technology with AI-driven data analytics. Similarly, AI's ability to provide deep insights into customer behavior and market trends is helping businesses tailor their offerings more effectively, enhancing customer satisfaction and loyalty.

Carl Solder highlights how AI is transforming business operations, making them more efficient and competitive. Predictive analytics, for example, can forecast market changes or customer needs, allowing companies to react more swiftly and strategically. Additionally, AI-driven tools are improving the quality and efficiency of customer service, providing personalized experiences that were previously costly and difficult to scale.

Societal Transformations

Beyond economic implications, AI is also affecting societal structures. It has the potential to greatly improve quality of life by making services more accessible and affordable. For example, AI in health care can help diagnose diseases earlier and with greater accuracy, potentially saving lives and reducing healthcare costs.

However, the societal impact of AI also presents challenges. The shift in job requirements and the potential for increased inequality are significant concerns. There is a need for policies and educational programs that can help the workforce transition to the AI-driven economy, ensuring that the benefits of AI are distributed equitably across society.

The economic and societal impacts of AI are far-reaching, affecting every corner of the globe. As businesses continue to harness the power of AI, they must also address the challenges it presents, ensuring that its benefits are realized while minimizing its risks. The future will likely see AI as a standard component of business and society, integral to driving innovation, economic growth, and social well-being.

The Future Is Now

As we stand on the brink of a technological revolution driven by artificial intelligence, the future of business and society is being rewritten. The insights provided by experts like James Larmer and Carl Solder underscore the transformative power of AI across various sectors, highlighting both the opportunities and challenges it presents.

Reinventing Business and Society

AI is not merely a tool for operational efficiency; it is a catalyst for innovation. By personalizing interactions, enhancing product capabilities, and redefining the interface between people and devices, AI is enabling businesses to deliver unparalleled customer experiences and develop new markets. Moreover, the economic impact of AI, as projected by experts, suggests significant contributions to global GDP, with new jobs created, and industries transformed.

Navigational Challenges

However, this transformation comes with substantial challenges. The displacement of jobs by automation, the ethical considerations of AI deployment, and the societal shifts due to technological disruption require thoughtful navigation. Businesses and policymakers must collaborate to create frameworks that not only foster innovation but also address the potential negative impacts of AI, ensuring that the transition to an AI-driven world is equitable and inclusive.

Future Directions

Looking forward, the integration of AI into daily life and business operations will continue to accelerate. Companies that can adapt to this new landscape, leveraging AI to enhance their offerings and streamline their operations, will likely emerge as leaders in the new economy. Education systems and workforce training programs will also need to evolve, preparing individuals for the new types of jobs created by AI advancements.

The role of AI in building and sustaining economic and societal growth cannot be overstated. As AI technologies continue to evolve and permeate more aspects of our lives, they promise not only to enhance productivity and efficiency but also to improve the quality of life for people around the world.

It is a pivotal moment for all stakeholders involved—business leaders, policymakers, educators, and the public—to engage with AI responsibly. By embracing AI's potential while carefully managing its challenges, we can harness its power to create a more prosperous, equitable, and innovative future.

Finally, thank you for reading my book. The journey ahead is complex and full of unknowns, but with careful planning and collaboration, the future shaped by AI holds incredible promise for us all.

Index

© Jonas Bjerg 2024
J. Bjerg, *The Early-Career Professional's Guide to Generative AI*,
https://doi.org/10.1007/979-8-8688-0456-4